ALSO BY ANTONIO DAMASIO

Descartes' Error: Emotion, Reason, and the Human Brain

The Feeling of What Happens:
Body and Emotion in the Making of Consciousness

Looking for Spinoza: Joy, Sorrow, and the Feeling Brain

Self Comes to Mind

Self
Comes to Mind

CONSTRUCTING THE
CONSCIOUS BRAIN

Antonio Damasio

PANTHEON BOOKS, NEW YORK

All rights reserved. Published in the United States by Pantheon Books,
a division of Random House, Inc., New York, and in Canada
by Random House of Canada Limited, Toronto.

Pantheon Books and colophon are registered trademarks of
Penguin Random House, LLC.

Library of Congress Cataloging-in-Publication Data
Damasio, Antonio R.
Self comes to mind : constructing the conscious brain / Antonio Damasio.
p.   cm.
Includes bibliographical references and index.
ISBN 978-0-307-37875-0
1. Brain—Evolution.   2. Developmental neurobiology.
3. Consciousness.   I. Title.
QP376.D356 2010
612.8'23—dc22   2010011660

Proprietary ISBN 978-0-525-61646-7

www.pantheonbooks.com
Book design by Virginia Tan
Printed in the United States of America
First Edition
2 4 6 8 9 7 5 3 1

For Hanna

My soul is like a hidden orchestra; I do not know which instruments grind and play away inside of me, strings and harps, timbales and drums. I can only recognize myself as symphony.

—FERNANDO PESSOA, *The Book of Disquiet*

What I cannot build, I cannot understand.

—RICHARD FEYNMAN

CONTENTS

PART I

Starting Over

I

Awakening

When I woke up, we were descending. I had been asleep long enough to miss the announcements about the landing and the weather. I had not been aware of myself or my surroundings. I had been unconscious.

Few things about our biology are as seemingly trivial as this commodity known as consciousness, the phenomenal ability that consists of having a mind equipped with an owner, a protagonist for one's existence, a self inspecting the world inside and around, an agent seemingly ready for action.

Consciousness is not merely wakefulness. When I woke up, two brief paragraphs ago, I did not look around vacantly, taking in the sights and the sounds as if my awake mind belonged to no one. On the contrary, I knew, almost instantly, with little hesitation if any, without effort, that this was me, sitting on an airplane, my flying identity coming home to Los Angeles with a long to-do list before the day would be over, aware of an odd combination of travel fatigue and enthusiasm for what was ahead, curious about the runway we would be landing on, and attentive to the adjustments of engine power that were bringing us to earth. No doubt, being awake was indispensable to this state, but wakefulness was hardly its main feature. What was that main feature? The fact that the myriad contents displayed in my mind, regardless of how vivid or well ordered, *connected* with me, the proprietor of my mind, through invisible strings that brought those contents together in the forward-moving feast we call self; and, no less important, the fact that the connection was *felt*. There was a *feelingness* to the experience of the connected me.

Awakening meant having my temporarily absent mind returned, but with *me* in it, both property (the mind) and proprietor (me) accounted for. Awakening allowed me to reemerge and survey my mental domains, the sky-wide projection of a magic movie, part documentary and part fiction, otherwise known as the conscious human mind.

We all have free access to consciousness, bubbling so easily and abundantly in our minds that without hesitation or apprehension we let it be turned off every night when we go to sleep and allow it to return every morning when the alarm clock rings, at least 365 times a year, not counting naps. And yet few things about our beings are as remarkable, foundational, and seemingly mysterious as consciousness. Without consciousness—that is, a mind endowed with subjectivity—you would have no way of knowing that you exist, let alone know who you are and what you think. Had subjectivity not begun, even if very modestly at first, in living creatures far simpler than we are, memory and reasoning are not likely to have expanded in the prodigious way they did, and the evolutionary road for language and the elaborate human version of consciousness we now possess would not have been paved. Creativity would not have flourished. There would have been no song, no painting, and no literature. Love would never have been love, just sex. Friendship would have been mere cooperative convenience. Pain would never have become suffering—not a bad thing, come to think of it— but an equivocal advantage given that pleasure would not have become bliss either. Had subjectivity not made its radical appearance, there would have been no knowing and no one to take notice, and consequently there would have been no history of what creatures did through the ages, no culture at all.

Although I have not yet provided a working definition of consciousness, I hope I am leaving no doubt as to what it means *not* to have consciousness: in the absence of consciousness, the personal view is suspended; we do not know of our existence; and we do not know that anything else exists. If consciousness had not developed in the course of evolution and expanded to its human version, the humanity we are now familiar with, in all its frailty and strength, would never have developed either. One shudders to think that a simple turn not taken might have meant the loss of the biological alternatives that make us truly human. But then, how would we ever have found out that something was missing?

We take consciousness for granted because it is so available, so easy to use, so elegant in its daily disappearing and reappearing acts, and yet, when we think of it, scientists and nonscientists alike, we do puzzle. What is consciousness made of? Mind with a twist, it seems to me, since we cannot be conscious without having a mind to be conscious of. But what is mind made of? Does mind come from the air or from the body? Smart people say it comes from the brain, that it *is* in the brain, but that is not a satisfactory reply. How does the brain *do* mind?

The fact that no one sees the minds of others, conscious or not, is especially mysterious. We can observe their bodies and their actions, what they do or say or write, and we can make informed guesses about what they think. But we cannot observe their minds, and only we ourselves can observe ours, from the inside, and through a rather narrow window. The properties of minds, let alone conscious minds, appear to be so radically different from those of visible living matter that thoughtful folk wonder how one process (conscious minds working) meshes with the other process (physical cells living together in aggregates called tissues).

But to say that conscious minds are mysterious—and on the face of it they are—is different from saying that the mystery is insoluble. It is different from saying that we shall never be able to understand how a living organism endowed with a brain develops a conscious mind.[1]

Goals and Reasons

This book is dedicated to addressing two questions. First: how does the brain construct a mind? Second: how does the brain make that mind conscious? I am well aware that addressing questions is not the same as answering them, and that on the matter of the conscious mind, it would be foolish to presume definitive answers. Moreover, I realize that the study of consciousness has expanded so much that it is no longer possible to do justice to all contributions being made to it. That, along with issues of terminology and perspective, make current work on consciousness resemble a walk through a minefield. Nonetheless, at one's own peril, it is reasonable to think through the questions and use the current evidence, incomplete and provisional as it is, to build testable conjectures and dream about the future. The goal of this book is to reflect on the conjectures and discuss a framework of hypotheses. The focus is on how the human brain needs to be structured and how it needs to operate in order for conscious minds to emerge.

Books should be written for a reason, and this one was written to start over. I have been studying the human mind and brain for more than thirty years, and I have previously written about consciousness in scientific articles and books.[2] But I have grown dissatisfied with my account of the problem, and reflection on relevant research findings, new and old, has changed my views, on two issues in particular: the origin and nature of feelings and the mechanisms behind the construction of the self. This book is an attempt to discuss the current views. In no small measure, the book is also about what we still do not know but wish we did.

The remainder of Chapter 1 situates the problem, explains the framework chosen to address it, and previews the main ideas that will emerge in the chapters ahead. Some readers may find that the long presentation in Chapter 1 slows down the reading, but I promise it will also make the rest of the book all the more accessible.

Approaching the Problem

Before we attempt to make some headway on the matter of how the human brain constructs a conscious mind, we need to acknowledge two important legacies. One of them consists of prior attempts to discover the neural basis of consciousness, in efforts that date back to the middle of the twentieth century. In a series of pioneering studies conducted in North America and Italy, a small band of investigators pointed with astonishing certainty to a brain sector that is now unequivocally related to the making of consciousness—the brain stem—and identified it as a critical contributor to consciousness. Not surprisingly, in light of what we know today, the account provided by these pioneers—Wilder Penfield, Herbert Jasper, Giuseppe Moruzzi, and Horace Magoun—was incomplete, and parts of it were less than correct. But one should have nothing but praise and admiration for the scientists who intuited the right target and aimed at it with such precision. This was the brave beginning of the enterprise to which several of us wish to contribute today.[3]

Also part of this legacy are studies performed more recently in neurological patients whose consciousness was compromised by focal brain damage. The work of Fred Plum and Jerome Posner launched the effort.[4] Over the years these studies, complementing those of the consciousness-research pioneers, have yielded a powerful collection of facts regarding the brain structures that are or are not involved in making human minds conscious. We can build on that foundation.

The other legacy to be acknowledged consists of a long tradition of formulating conceptions of mind and consciousness. It has a rich history, as long and varied as the history of philosophy. From the wealth of its offerings, I have come to favor the writings of William James as an anchor for my own thinking, although this does not imply a full endorsement of his positions on consciousness and especially on feeling.[5]

The title of this book, as well as its first pages, leave no doubt that in approaching the conscious mind, I privilege the self. I believe conscious minds arise when a self process is added onto a basic mind process. When selves do not occur within minds, those minds are not conscious in the proper sense. This is a predicament faced by humans whose self process is suspended by dreamless sleep, anesthesia, or brain disease.

Defining the self process that I regard as so indispensable for consciousness, however, is easier said than done. That is why William James is so helpful to this preamble. James wrote eloquently about the importance of the self, and yet he also noted that, on many occasions, the presence of the self is so subtle that the contents of the mind dominate consciousness as they stream along. We need to confront this elusiveness and decide on its consequences before we go any further. Is there a self, or is there not? If there is a self, is it present whenever we are conscious, or is it not?

The answers are unequivocal. There is indeed a self, but it is a process, not a thing, and the process is present at all times when we are presumed to be conscious. We can consider the self process from two vantage points. One is the vantage point of an observer appreciating a dynamic *object*—the dynamic object constituted by certain workings of minds, certain traits of behavior, and a certain history of life. The other vantage point is that of the self as *knower,* the process that gives a focus to our experiences and eventually lets us reflect on those experiences. Combining the two vantage points produces the dual notion of self used throughout the book. As we shall see, the two notions correspond to two stages of evolutionary development of the self, the self-as-knower having had its origin in the self-as-object. In everyday life each notion corresponds to a level of operation of the conscious mind, the self-as-object being simpler in scope than the self-as-knower.

From either vantage point, the process has varied scopes and intensities and its manifestations vary with the occasions. The self can operate on a subtle register, as "a hint half hinted" of the presence of a living organism,[6] or on a salient register that includes personhood and identity for the owner of the mind. Now you sense it, now you don't, but you always *feel* it, is my way of summing up the situation.

James thought that the self-as-object, the material me, was the sum total of all that a man could call his—"not only his body and his psychic powers, but his clothes and his wife and children, his ancestors and friends, his reputation and works, his lands and horses, and yacht and bank account."[7] Leaving aside the political incorrectness, I agree. But James also thought something else with which I am in even greater agreement: what allows the mind to know that such dominions exist and belong to their mental owners—body, mind, past and present, and all the rest—is that the perception of any of these items generates emotions and feelings, and, in turn, the feelings accomplish the separation between the contents that belong to the self and those that do not. From my perspective, such feelings operate as *markers*. They are the emotion-based signals I designate as somatic markers.[8] When contents that pertain to the self occur in the mind stream, they provoke the appearance of a marker, which joins the mind stream as an image, juxtaposed to the image that prompted it. These feelings accomplish a distinction between self and nonself. They are, in a nutshell, *feelings of knowing*. We shall see that the construction of a conscious mind depends, at several stages, on the generation of such feelings. As for my working definition of the material me, the self-as-object, it is as follows: *a dynamic collection of integrated neural processes, centered on the representation of the living body, that finds expression in a dynamic collection of integrated mental processes.*

The self-as-subject, as knower, as the "I," is a more elusive presence, far less collected in mental or biological terms than the *me*, more dispersed, often dissolved in the stream of consciousness, at times so annoyingly subtle that it is there but almost not there. The self-as-knower is more difficult to capture than the plain me, unquestionably. But that does not diminish its significance for consciousness. The self-as-subject-and-knower is not only a very real presence but a turning point in biological evolution. We can imagine that the self-as-subject-and-knower is stacked, so to speak, on top of the self-as-object, as a new layer of neural processes giving rise to yet another layer of mental processing. There is no dichotomy between self-as-object and self-as-knower; there is, rather, a continuity and progression. The self-as-knower is grounded on the self-as-object.

Consciousness is not merely about images in the mind. It is, in the very least, about an *organization of mind contents centered on the organism that produces and motivates those contents.* But consciousness, in the sense that reader and author can experience anytime they wish, is more than a mind organized under the influence of a living, acting organism. It is also a mind capable of knowing that such a living, acting organism exists. To be sure, the fact that the brain succeeds in creating neural patterns that map things experienced as images is an important part of the process of being conscious. Orienting the images in the perspective of the organism is also a part of the process. But that is not the same as automatically and explicitly *knowing* that images exist within me and are mine and, in current lingo, actionable. The mere presence of organized images flowing in a mental stream produces a mind, but unless some supplementary process is added on, the mind remains *unconscious.* What is missing from that unconscious mind is a *self.* What the brain needs in order to become conscious is to acquire a new property— *subjectivity*—and a defining trait of subjectivity is the feeling that pervades the images we experience subjectively. For a contemporary treatment of the importance of subjectivity from the perspective of philosophy, read John Searle's *The Mystery of Consciousness.*[9]

In keeping with this idea, the decisive step in the making of consciousness is not the making of images and creating the basics of a mind. The decisive step is *making the images ours,* making them belong to their rightful owners, the singular, perfectly bounded organisms in which they emerge. In the perspective of evolution and in the perspective of one's life history, the knower came in steps: the protoself and its primordial feelings; the action-driven core self; and finally the autobiographical self, which incorporates social and spiritual dimensions. But these are dynamic processes, not rigid things, and on any day their level fluctuates (simple, complex, somewhere in between) and can be readily adjusted as the circumstances dictate. A knower, by whatever name one may want to call it—self, experiencer, protagonist—needs to be generated in the brain if the mind is to become conscious. When the brain manages to introduce a knower in the mind, subjectivity follows.

Should the reader wonder if this defense of the self is necessary, let me say that it is quite justified. At this very moment, those of us in neuroscience whose work aims at elucidating consciousness subscribe to very different attitudes toward the self. The attitudes range from considering the self as an indispensable topic of the research agenda to thinking that the time has not come to deal with the subject (literally!).[10] Given that the work associated with either attitude continues to produce useful ideas, there is no need, as yet, to decide which approach will turn out to be more satisfactory. But we need to acknowledge that the resulting accounts are different.

In the meantime, it is noteworthy that these two attitudes perpetuate a difference of interpretation that separated William James from David Hume, one that is generally overlooked in such discussions. James wanted to make certain that his conceptions of self had a firm biological grounding: his "self" would not be mistaken for a metaphysical knowing agency. But that did not prevent him from recognizing a knowing function for the self, even when the function was subtle rather than exuberant. David Hume, on the other hand, pulverized the self to the point of doing away with it. The following passages illustrate Hume's views: "I never can catch *myself* at any time without a perception and never can observe anything but the perception." And further on: "I may venture to affirm of the rest of mankind, that they are nothing but a bundle or collection of different perceptions, which succeed each other with an inconceivable rapidity, and are in a perceptual flux and movement."

Commenting on Hume's dismissal of the self, James was moved to issue a memorable rebuke and affirm the existence of the self, emphasizing the odd mixture of "unity and diversity" within it and calling attention to the "core of sameness" running through the ingredients of the self.[11]

The foundation discussed here has been modified and expanded upon by philosophers and neuroscientists to include different aspects of self.[12] But the significance of the self for the construction of the conscious mind has not been diminished. I doubt that the neural basis for the conscious mind can be comprehensively elucidated without first accounting for the self-as-object—the material me—and for the self-as-knower.

Contemporary work on philosophy of mind and psychology has extended the conceptual legacy, while the extraordinary development of general biology, evolutionary biology, and neuroscience has capitalized on the neural legacy, produced a wide array of techniques to investigate the brain, and amassed a colossal amount of facts. The evidence, conjectures, and hypotheses presented in this book are grounded on all these developments.

The Self as Witness

Countless creatures for millions of years have had active minds, but only in those who developed a self capable of operating as a witness to the mind was its existence acknowledged, and only after minds developed language and lived to tell did it become widely known that minds did exist. The self as witness is the something extra that reveals the presence, in each of us, of events we call mental. We need to understand how that something extra is created.

The notions of witness and protagonist are not meant as mere literary metaphors. I hope they help illustrate the range of roles that the self assumes in the mind. For one thing, the metaphors can help us see the situation we face when we attempt to understand mental processes. A mind unwitnessed by a self protagonist is still a mind. However, given that the self is our only natural means to know the mind, we are entirely dependent on the self's presence, capabilities, and limits. And given this systematic dependence, it is extremely difficult to imagine the nature of the mind process independently of the self, although from an evolutionary perspective, it is apparent that plain mind processes preceded self processes. The self permits a view of the mind, but the view is clouded. The aspects of the self that permit us to formulate interpretations about our existence and about the world are still evolving, certainly at the cultural level and, in all likelihood, at the biological level as well. For instance, the upper reaches of self are still being modified by all manner of social and cultural interactions and by the accrual of scientific knowledge about the very workings of mind and brain. One entire century of movie viewing has certainly had an impact on the human self, as has the spectacle of globalized societies now instantly broadcast by electronic media. As for the impact of the digital revolution, it is just beginning to be appreciated. In brief, our only direct view of the mind depends on a part of that very mind, a self process that we have good reason to believe cannot provide a comprehensive and reliable account of what is going on.

At first glance, after acknowledging the self as our entry into knowledge, it may appear paradoxical, not to mention ungrateful, to question its reliability. And yet that is the situation. Except for the direct window that the self opens into our pains and pleasures, the information it provides must be questioned, most certainly when the information pertains to its very nature. The good news, however, is that the self also has made reason and scientific observation possible, and reason and science, in turn, have been gradually correcting the misleading intuitions prompted by the unaided self.

Overcoming a Misleading Intuition

It is arguable that cultures and civilizations would not have come to pass in the absence of consciousness, thus making consciousness a notable development in biological evolution. And yet the very nature of consciousness poses serious problems for those attempting to elucidate its biology. Viewing consciousness from where we stand today, mindful and armed with a self, can be blamed for an understandable but troubling distortion of the history of mind and consciousness studies. Viewed from the top, the mind acquires a special status, discontinuous with the remainder of the organism to which it belongs. Viewed from the top, the mind appears to be not just very complex, which it certainly is, but also different in kind from the biological tissues and functions of the organism that begets it. In practice, we adopt two sorts of optic when we observe our beings: we see the mind with eyes that are turned inward; and we see biological tissues with eyes that are turned outward. (To boot, we use microscopes to extend our vision.) Under the circumstances, it is not surprising that the mind appears to have a nonphysical nature and that its phenomena appear to belong to another category.

Viewing the mind as a nonphysical phenomenon, discontinuous with the biology that creates and sustains it, is responsible for placing the mind outside the laws of physics, a discrimination to which other brain phenomena are not usually subject. The most striking manifestation of this oddity is the attempt to connect the conscious mind to heretofore undescribed properties of matter and, for example, explain consciousness in terms of quantic phenomena. The rationale for this idea appears to be as follows: the conscious mind seems mysterious; because quantum physics remains mysterious, perhaps the two mysteries are connected.[13]

Given our incomplete knowledge of both biology and physics, one should be cautious before dismissing alternative accounts. After all, in spite of neurobiology's remarkable success, our understanding of the human brain is quite incomplete. Nonetheless, the possibility of explaining mind and consciousness parsimoniously, within the confines of neurobiology as currently conceived, remains open; it should not be abandoned unless the technical and theoretical resources of neurobiology are exhausted, an unlikely prospect at the moment.

Our intuition tells us that the mercurial, fleeting business of the mind lacks physical extension. I believe this intuition is false and attributable to the limitations of the unaided self. I see no reason to give to it more credence than to previously evident and powerful intuitions such as the pre-Copernican view of what the sun does to the earth or, for that matter, the view that the mind resides in the heart. Things are not always what they seem. White light is a composite of the colors of the rainbow, although that is not apparent to the naked eye.[14]

An Integrated Perspective

Most of the progress made to date on the neurobiology of conscious minds has been based on combining three perspectives: (1) the direct-witness perspective on the individual conscious mind, which is personal, private, and unique to each one of us; (2) the behavioral perspective, which allows us to observe the telltale actions of others whom we have reason to believe also have conscious minds; and (3) the brain perspective, which allows us to study certain aspects of brain function in individuals whose conscious mind states are presumed to be either present or absent. Evidence from these three perspectives, even when intelligently aligned, is usually not enough to generate a smooth transition across the three kinds of phenomena—introspective, first-person inspection; external behaviors; and brain events. In particular, there appears to be a major gap between the evidence from first-person introspection and the evidence from brain events. How can we bridge such gaps?

A fourth perspective is needed, one that requires a radical change in the way the history of conscious minds is viewed and told. In earlier work I advanced the idea of turning life regulation into the support and justification of self and consciousness, and that idea suggested a path into this new perspective: a search for antecedents of self and consciousness in the evolutionary past.[15] Accordingly, the fourth perspective is grounded on facts from evolutionary biology and neurobiology. It requires us to consider early living organisms first, then gradually move across evolutionary history toward current organisms. It requires us to note incremental modifications of nervous systems and link them to the incremental emergence of, respectively, behavior, mind, and self. It also requires an internal working hypothesis: that mental events are equivalent to certain kinds of brain events. Of course, mental activity is caused by the brain events that antecede it, but at the end of the day, the mental events correspond to certain states of brain circuits. In other words, some neural patterns *are* simultaneously mental images. When some other neural patterns generate a rich enough self process subject, the images can become *known*. But if no self is generated, the images still *are,* although no one, inside or outside the organism, knows of their existence. Subjectivity is not required for mental states to exist, only for them to be privately known.

In brief, the fourth perspective asks us to construct, simultaneously, with the help of available facts, a view from the past, and from within, literally an imagined view of a brain caught in the state of containing a conscious mind. To be sure, this is a conjectural, hypothetical view. There are facts to support parts of this imaginarium, but it is in the nature of the "mind-self-body-brain problem" that we must live for quite a while with theoretical approximations rather than complete explanations.

It might be tempting to regard the hypothesized equivalence of mind events to certain brain events as a crude reduction of the complex to the simple. This would be a false impression, however, given that neurobiological phenomena are immensely complex to begin with, anything but simple. The explanatory reductions involved here are not from the complex to the simple but rather from the extremely complex to the slightly less so. Although this book is not about the biology of simple organisms, the facts to which I allude in Chapter 2 make it clear that the lives of cells occur in extraordinary complex universes that formally resemble, in many ways, our elaborate human universe. The world and behavior of a single-cell organism such as the paramecium are a wonder to behold, far closer to who we are than meets the eye.

It is also tempting to interpret the proposed brain-mind equivalence as a neglect of the role of culture in the generation of the mind or as a downgrading of the role of individual effort in the shaping of the mind. Nothing could be farther from my formulation, as will become clear.

Using the fourth perspective, I can now rephrase some of the statements made earlier in a way that takes into account facts from evolutionary biology and includes the brain: countless creatures for millions of years have had active minds happening in their *brains,* but only after those *brains* developed a protagonist capable of bearing witness did consciousness begin, in the strict sense, and only after those *brains* developed language did it become widely known that minds did exist. The witness is the something extra that reveals the presence of implicit *brain* events we call mental. Understanding how the brain produces that something extra, the protagonist we carry around and call self, or me, or I, is an important goal of the neurobiology of consciousness.

The Framework

Before I sketch the framework guiding this book, I need to introduce some basic facts. Organisms make minds out of the activity of special cells known as neurons. Neurons share most of the characteristics of other cells in our body, and yet their operation is distinctive. They are sensitive to changes around them; they are excitable (an interesting property they share with muscle cells). Thanks to a fibrous prolongation known as the axon, and to the end region of the axon known as the synapse, neurons can send signals to other cells—other neurons, muscle cells—often quite far away. Neurons are largely concentrated in a central nervous system (the brain, for short), but they send signals to the organism's body, as well as to the outside world, and they receive signals from both.

The number of neurons in each human brain is on the order of billions, and the synaptic contacts that the neurons make among themselves number in the trillions. Neurons are organized in small microscopic circuits, whose combination constitutes progressively larger circuits, which in turn form networks or systems. For more on neurons and brain organization, see Chapter 2 and the Appendix.

Minds emerge when the activity of small circuits is organized across large networks so as to compose momentary patterns. The patterns represent things and events located outside the brain, either in the body or in the external world, but some patterns also represent the brain's own processing of other patterns. The term *map* applies to all those representational patterns, some of which are coarse, while others are very refined, some concrete, others abstract. In brief, the brain maps the world around it and maps its own doings. Those maps are experienced as *images* in our minds, and the term *image* refers not just to the visual kind but to images of any sense origin such as auditory, visceral, tactile, and so forth.

Let us now turn to the framework proper. Using the term *theory* to describe proposals for how the brain produces this or that phenomenon is somewhat out of place. Unless the scale is large enough, most theories are just hypotheses. What is being proposed in this book, however, is more than that, since it articulates several hypothetical components for one aspect or another of the phenomena I am addressing. What we hope to explain is too complex to be addressed by a single hypothesis and be accounted for by one mechanism. So I have settled for the term *framework* to designate the effort.

In order to qualify for the lofty title, the ideas presented in the chapters ahead need to accomplish certain goals. Given that we wish to understand how the brain makes the mind conscious, and given that it is manifestly impossible to deal with all levels of brain function in assembling an explanation, the framework must specify the level at which the explanation applies. This is the large-scale systems level, the level at which macroscopic brain regions constituted by neuron circuits interact with other such regions to form systems. Of necessity, those systems are macroscopic, but the underlying microscopic anatomy is known in part, as are the general operating rules of the neurons that constitute them. The large-scale systems level is amenable to research via numerous techniques, old and new. They include the modern version of the lesion method (which relies on the study of neurological patients with focal brain damage investigated with structural neuroimaging and experimental cognitive and neuropsychological techniques); functional neuroimaging (based on magnetic resonance scanning, positron-emission tomography, magnetoencephalography, and assorted electrophysiological techniques); direct neurophysiological recording of neuron activity in the setting of neurosurgical treatments; and transcranial magnetic stimulation.

The framework must interconnect behavior, mind, and brain events. On this second goal, the framework aligns behavior, mind, and brain closely; and because it relies on evolutionary biology, it places consciousness in a historical setting, a placement suitable to organisms undergoing evolutionary transformation by natural selection. Moreover, the maturation of neuron circuitries in each brain is also seen as subject to selection pressures resulting from the very activity of organisms and the processes of learning. The repertoires of neuron circuitries initially provided by the genome are changed accordingly.[16]

The framework indicates the placement of regions involved in mind-making, at whole-brain scale, and proposes how some brain regions might operate in concert to produce the self. It suggests how a brain architecture that features convergence and divergence of neuron circuitries plays a role in the high-order coordination of images and is essential for the construction of the self and of other aspects of mental function, namely memory, imagination, language, and creativity.

The framework needs to break down the phenomenon of consciousness in components amenable to neuroscience research. The result is two researchable domains, namely, mind processes and self processes. Furthermore, it decomposes the self process into subtypes. The latter separation offers two advantages: presuming and investigating consciousness in species that are likely to have self processes albeit less elaborate; and creating a bridge between the high levels of self and the sociocultural space in which humans operate.

Another goal: the framework must address the issue of how system macroevents are built from microevents. Here the framework hypothesizes the equivalence of mental states to certain states of regional brain activity. The framework assumes that when certain ranges of intensity and frequency of neuron firing occur in small neuron circuits, when some of these circuits are synchronously activated, and when certain conditions of network connectivity are met, the result is a "mind with feelings." In other words, as a result of the growing size and complexity of neural networks, there is a scaling up of "cognition" and "feeling," from the microlevel to the macrolevel, across hierarchies. The model for this scaling up to mind with feeling can be found in the physiology of movement. The contraction of a single microscopic muscle cell is a negligible phenomenon, whereas the simultaneous contraction of large numbers of muscle cells can produce visible movement.

A Preview of Main Ideas

I

Of the ideas advanced in the book, none is more central than the notion that the body is a foundation of the conscious mind. We know that the most stable aspects of body function are represented in the brain, in the form of maps, thereby contributing images to the mind. This is the basis of the hypothesis that the special kind of mental images of the body produced in body-mapping structures, constitutes the *protoself*, which foreshadows the self to be. Of note, the critical body-mapping and image-making structures are located below the level of the cerebral cortex, in a region known as the upper brain stem. This is an old part of the brain shared with many other species.

II

Another central idea is based on the consistently overlooked fact that the brain's protoself structures are not merely *about* the body. They are literally and inextricably *attached* to the body. Specifically, they are attached to the parts of the body that bombard the brain with their signals, at all times, only to be bombarded back by the brain and, by so doing, creating a resonant loop. This resonant loop is perpetual, broken only by brain disease or death. Body and brain *bond*. As a result of this arrangement, the protoself structures have a privileged and direct relationship to the body. The images they engender regarding the body are conceived in circumstances different from those of other brain images, say, visual or auditory. In light of these facts, the body is best conceived as the rock on which the protoself is built, while the protoself is the pivot around which the conscious mind turns.

III

I hypothesize that the first and most elementary product of the protoself is *primordial feelings,* which occur spontaneously and continuously whenever one is awake. They provide a direct experience of one's own living body, wordless, unadorned, and connected to nothing but sheer existence. These *primordial feelings* reflect the current state of the body along varied dimensions, for example, along the scale that ranges from pleasure to pain, and they originate at the level of the brain stem rather than the cerebral cortex. All feelings of emotion are complex musical variations on primordial feelings.[17]

In the functional arrangement outlined here, pain and pleasure are body events. The events are *also* mapped in a brain that at no instant is separated from its body. Thus primordial feelings are a special kind of image generated thanks to the obligate body-brain interaction, to the characteristics of the circuitry accomplishing the connection, and possibly to certain properties of neurons. It is not enough to say that feelings are felt because they map the body. I hypothesize that in addition to holding a unique relationship to the body, the brain-stem machinery responsible for making the kinds of images we call feelings is capable of richly mixing signals from the body and thus creating complex states with the special and novel properties of feeling rather than mere slavish maps of the body. The reason why nonfeeling images are also felt is that they are normally *accompanied* by feelings.

The foregoing implies that the notion of a sharp border separating body and brain is problematic. It also suggests a potentially fruitful approach to the vexing problem of why and how normal mental states are invariably imbued with some form of feeling.

IV

Brains begin building conscious minds not at the level of the cerebral cortex but rather at the level of the brain stem. Primordial feelings are not only the first images generated by the brain but also immediate manifestations of sentience. They are the protoself foundation for more complex levels of self. These ideas run counter to widely accepted views, although Jaak Panksepp (cited earlier) has defended a comparable position and so has Rodolfo Llinás. But the conscious mind as we know it is a far different affair from the conscious mind that emerges in the brain stem, and on this point there probably is universal agreement. The cerebral cortices endow the mind-making process with a profusion of images that, as Hamlet might put it, go far beyond anything that poor Horatio could ever dream of, in heaven or earth.

Conscious minds begin when self comes to mind, when brains add a self process to the mind mix, modestly at first but quite robustly later. The self is built in distinct steps grounded on the *protoself*. The first step is the generation of primordial feelings, the elementary feelings of existence that spring spontaneously from the protoself. Next is the *core self*. The core self is about action—specifically, about a relationship between the organism and the object. The core self unfolds in a sequence of images that describe an object engaging the protoself and modifying that protoself, including its primordial feelings. Finally, there is the *autobiographical self*. This self is defined in terms of biographical knowledge pertaining to the past as well as the anticipated future. The multiple images whose ensemble defines a biography generate pulses of core self whose aggregate constitutes an autobiographical self.

The protoself with its primordial feelings, and the core self, constitute a "material me." The autobiographical self, whose higher reaches embrace all aspects of one's social persona, constitute a "social me" and a "spiritual me." We can observe these aspects of self within our own minds or study their effects in the behavior of others. In addition, however, the core and autobiographical selves within our minds construct a knower; in other words, they endow our minds with another variety of subjectivity. For practical purposes, normal human consciousness corresponds to a mind process in which all of these self levels operate, offering to a limited number of mind contents a momentary link to a pulse of core self.

V

At neither modest nor robust levels do self and consciousness *happen* in one area or region or center of the brain. Conscious minds result from the smoothly articulated operation of several, often many, brain sites. The key brain structures in charge of implementing the requisite functional steps include specific sectors of the upper brain stem, a set of nuclei in a region known as the thalamus, and specific but widespread regions of the cerebral cortex.

The ultimate consciousness product occurs *from* those numerous brain sites at the same time and not in one site in particular, much as the performance of a symphonic piece does not come from the work of a single musician or even from a whole section of an orchestra. The oddest thing about the upper reaches of a consciousness performance is the conspicuous absence of a conductor *before* the performance begins, although, as the performance unfolds, a conductor comes into being. For all intents and purposes, a conductor is now leading the orchestra, although the performance has created the conductor—the self—not the other way around. The conductor is cobbled together by feelings and by a narrative brain device, although this fact does not make the conductor any less real. The conductor undeniably exists in our minds, and nothing is gained by dismissing it as an illusion.

The coordination on which conscious minds depend is achieved by a variety of means. At the modest core level, it begins quietly, as a spontaneous assembly of images that emerge one after the other in close time proximity, the image of an object, on the one hand, and the image of the protoself changed by the object, on the other. No additional brain structures are needed for a core self to emerge, at this simple level. The coordination is natural, sometimes resembling a mere musical duo, played by organism and object, sometimes resembling a chamber music ensemble, and in both cases managing quite well without a conductor. But when the contents being processed in the mind are more numerous, other devices are required to accomplish coordination. In that case a variety of brain regions below the level of the cerebral cortices and within them play a key role.

Building a mind capable of encompassing one's lived past and anticipated future, along with the lives of others added to the fabric and a capacity for reflection to boot, resembles the execution of a symphony of Mahlerian proportions. But the marvel, as hinted at earlier, is that the score and the conductor become reality only as life unfolds. The coordinators are not mythical, sapient homunculi in charge of interpreting anything. And yet the coordinators do help with the assembly of an extraordinary media universe and with the placement of a protagonist in its midst.

The grand symphonic piece that is consciousness encompasses the foundational contributions of the brain stem, forever hitched to the body, and the wider-than-the-sky imagery created in the cooperation of cerebral cortex and subcortical structures, all harmoniously stitched together, in ceaseless forward motion, interruptible only by sleep, anesthesia, brain dysfunction, or death.

No single mechanism explains consciousness in the brain, no single device, no single region, or feature, or trick, any more than a symphony can be played by one musician or even a few. Many are needed. What each of them contributes does count. But only the ensemble produces the result we seek to explain.

VI

Managing and safekeeping life efficiently are two of the recognizable achievements of consciousness: neurological patients whose consciousness is compromised are unable to manage their lives independently even when their basic life functions operate normally. And yet mechanisms for managing and maintaining life are not a novelty in biological evolution and are not necessarily dependent on consciousness. Such mechanisms already exist in single cells and are coded in their genome. They are also widely replicated within ancient, humble, *un*-minded and *un*-conscious neuron circuits, and they are very much present deep in human brains. We shall see that managing and safekeeping life is the fundamental premise of biological value. Biological value has influenced the evolution of brain structures, and in any brain it influences almost every step of brain operations. It is expressed as simply as in the release of chemical molecules related to reward and punishment, or as elaborately as in our social emotions and in sophisticated reasoning. Biological value naturally guides and colors, so to speak, almost everything that happens inside our very minded, very conscious brains. Biological value has the status of a principle.

In brief, the conscious mind emerges within the history of life regulation. Life regulation, a dynamic process known as *homeostasis* for short, begins in unicellular living creatures, such as a bacterial cell or a simple amoeba, which do not have a brain but are capable of adaptive behavior. It progresses in individuals whose behavior is managed by simple brains, as is the case with worms, and it continues its march in individuals whose brains generate both behavior and mind (insects and fish being examples). I am ready to believe that whenever brains begin to generate primordial feelings—and that could be quite early in evolutionary history—organisms acquire an early form of sentience. From there on, an organized self process could develop and be added to the mind, thereby providing the beginning of elaborate conscious minds. Reptiles are contenders for this distinction, for example; birds make even stronger contenders; and mammals get the award and then some.

Most species whose brains generate a self do so at core level. Humans have both core self and autobiographical self. A number of mammals are likely to have both as well, namely wolves, our ape cousins, marine mammals and elephants, cats, and, of course, that off-the-scale species called the domestic dog.

VII

The march of mind progress does not end with the arrival of the modest levels of self. Throughout the evolution of mammals, especially primates, minds become ever more complex, memory and reasoning expanding notably, and the self processes enlarge their scope. The core self remains, but it is gradually surrounded by an autobiographical self, whose neural and mental natures are very different from those of the core self. We become able to use a part of our mind's operation to monitor the operation of other parts. The conscious minds of humans, armed with such complex selves and supported by even greater capabilities of memory, reasoning, and language, engender the instruments of culture and open the way into new means of homeostasis at the level of societies and culture. In an extraordinary leap, homeostasis acquires an extension into the sociocultural space. Justice systems, economic and political organizations, the arts, medicine, and technology are examples of the new devices of regulation.

The dramatic reduction of violence along with the increase in tolerance that has become so apparent in recent centuries would not have occurred without sociocultural homeostasis. Neither would the gradual transition from coercive power to the power of persuasion that hallmarks advanced social and political systems, their failures notwithstanding. The investigation of sociocultural homeostasis can be informed by psychology and neuroscience, but the native space of its phenomena is cultural. It is reasonable to describe those who study the rulings of the U.S. Supreme Court, the deliberations of the U.S. Congress, or the workings of financial institutions as engaging, indirectly, in studying the vagaries of sociocultural homeostasis.

Both basic homeostasis (which is nonconsciously guided) and sociocultural homeostasis (which is created and guided by reflective conscious minds) operate as curators of biological value. Basic and sociocultural varieties of homeostasis are separated by billions of years of evolution, and yet they promote the same goal—the survival of living organisms—albeit in different ecological niches. That goal is broadened, in the case of sociocultural homeostasis, to encompass the *deliberate* seeking of well-being. It goes without saying that the way in which human brains manage life requires both varieties of homeostasis in continuous interaction. But while the basic variety of homeostasis is an established inheritance, provided by everyone's genome, the sociocultural variety is a somewhat fragile work in progress, responsible for much of human drama, folly, and hope. The interaction between these two kinds of homeostasis is not confined to each individual. There is growing evidence that, over multiple generations, cultural developments lead to changes in the genome.

VIII

Viewing the conscious mind in the optic of evolution from simple lifeforms toward complex and hypercomplex organisms such as ours helps naturalize the mind and shows it to be the result of stepwise progressions of complexity within the biological idiom.

We can look at human consciousness and at the functions it made possible (language, expanded memory, reasoning, creativity, the whole edifice of culture) as the curators of value inside our modern, very minded, very social beings. And we can imagine a long umbilical cord that links the barely weaned, perennially dependent conscious mind to the depths of very elementary and very *un*-conscious regulators of the value principle.

The history of consciousness cannot be told in the conventional way. Consciousness came into being because of biological value, as a contributor to more effective value management. But consciousness did not *invent* biological value or the process of valuation. Eventually, in human minds, consciousness revealed biological value and allowed the development of new ways and means of managing it.

Life and the Conscious Mind

Is it reasonable to devote a book to the question of how brains make conscious minds? It is sensible to ask if understanding the brain work behind mind and self has any practical significance besides satisfying our curiosity about human nature. Does it make any difference in daily life? For many reasons, large and small, I think it does. Brain science and its explanations are not about to provide for all people the satisfaction that so many obtain from experiencing the arts or cultivating spiritual beliefs. But there certainly are other compensations.

Understanding the circumstances in which conscious minds emerged in the history of life, and specifically how they developed in human history, allows us to judge perhaps more wisely than before the quality of the knowledge and advice those conscious minds provide. Is the knowledge reliable? Is the advice sound? Do we gain from under-standing the mechanisms behind the minds that give us counsel?

Elucidating the neural mechanisms behind conscious minds reveals that our selves are not always sound and that they are not in control of every decision. But the facts also authorize us to reject the false impression that our ability to deliberate consciously is a myth. Elucidating conscious as well as nonconscious mind processes increases the possibility of fortifying our deliberative powers. The self opens the way for deliberation and for the adventure of science, two specific tools with which all the misleading guidance of the unaided self can be countered.

The time will come when the issue of human responsibility, in general moral terms as well as on matters of justice and its application, will take into account the evolving science of consciousness. Perhaps the time is now. Armed with reflexive deliberation and scientific tools, an understanding of the neural construction of conscious minds also adds a welcome dimension to the task of investigating the development and shaping of cultures, the ultimate product of collectives of conscious minds. As humans debate the benefits or perils of cultural trends, and of developments such as the digital revolution, it may help to be informed about how our flexible brains create consciousness. For example, will the progressive globalization of human consciousness brought on by the digital revolution retain the goals and principles of basic homeostasis, as current sociocultural homeostasis does? Or will it break away from its evolutionary umbilical cord, for better or worse?[18]

Naturalizing the conscious mind and planting it firmly in the brain does not diminish the role of culture in the construction of human beings, does not reduce human dignity, and does not mark the end of mystery and puzzlement. Cultures arise and evolve from collective efforts of human brains, over many generations, and some cultures even die in the process. They require brains that have already been shaped by prior cultural effects. The significance of cultures to the making of the modern human mind is not in question. Nor is the dignity of that human mind diminished by connecting it to the astonishing complexity and beauty to be found inside living cells and tissues. On the contrary, connecting personhood to biology is a ceaseless source of awe and respect for anything human. Last, naturalizing the mind may solve one mystery but only to raise the curtain on other mysteries quietly awaiting their turn.

Placing the construction of conscious minds in the history of biology and culture opens the way to reconciling traditional humanism and modern science, so that when neuroscience explores human experience into the strange worlds of brain physiology and genetics, human dignity is not only retained but reaffirmed.

F. Scott Fitzgerald wrote memorably, "His was a great sin who first invented consciousness." I can understand why he said so, but his condemnation is only half the story, appropriate for moments of discouragement with the imperfections of nature that conscious minds expose so nakedly. The other half of the story should be occupied with praise for such an invention as the enabler of all the creations and discoveries that trade loss and grief for joy and celebration. The emergence of consciousness opened the way to a life worth living. Understanding how it comes about can only strengthen that worth.[19]

Does knowing about how the brain works matter at all for how we live our lives? I believe it matters very much, all the more so if, besides knowing who we presently are, we care at all for what we may become.

2

From Life Regulation to Biological Value

The Implausibility of Reality

Mark Twain thought that the big difference between fiction and reality was that fiction had to be believable. Reality could afford to be implausible, but fiction could not. And so the narrative of mind and consciousness that I am presenting here does not conform to the requirements of fiction. It is actually counterintuitive. It upsets traditional human storytelling. It repeatedly denies long-held assumptions and not a few expectations. But none of this makes the account any less likely.

The notion that hidden underneath conscious minds there are unconscious mind processes is hardly news. This idea was first aired more than a century ago, when the public greeted it with some surprise, but today the notion is commonplace. What is not commonly appreciated, although it is well known, is that long before living creatures had minds, they exhibited efficient and adaptive behaviors that for all intents and purposes resemble those that arise in mindful, conscious creatures. Of necessity, those behaviors were *not* caused by minds, let alone consciousness. In brief, it is not just that conscious and nonconscious processes coexist but rather that nonconscious processes that are relevant to maintaining life can exist without their conscious partners.

As far as mind and consciousness are concerned, evolution has brought us different sorts of brains. There is the sort of brain that produces behavior but does not appear to have mind or consciousness; an example is the nervous system of *Aplysia californica,* the marine snail that became popular in the laboratory of the neurobiologist Eric Kandel. Another sort produces the whole range of phenomena—behavior, mind, and consciousness—human brains are the prime example, of course. And a third sort of brain clearly produces behavior, is likely to produce a mind, but whether it generates consciousness in the sense discussed here is not so clear. That is the case of insects.

But the surprises do not end with the notion that in the absence of mind and consciousness brains can produce respectable behaviors. It turns out that living creatures without any brain at all, down to single cells, exhibit seemingly intelligent and purposeful behavior as well. And that too is an underappreciated fact.

There is no doubt that we can gain useful insights into how human brains produce conscious minds by understanding the simpler brains that produce neither mind nor consciousness. As we engage in that retrospective survey, however, it becomes apparent that in order to explain the rise of such long-ago brains we need to go even deeper into the past, further back into the world of simple life-forms, devoid of both minds *and* brains, those life-forms that are unconscious, unminded, and *un*-brained. In fact, if we are to find out the rhymes and reasons behind conscious brains, we need to get closer to the beginnings of life. And here once again we come to notions that not only are surprising but undermine commonly held assumptions about the contributions of brains, minds, and consciousness to the management of life.

Natural Will

We need a fable again. Once upon a time life came about in the lengthy history of evolution. This was 3.8 billions of years ago, when the ancestor of all future organisms made its first appearance. Some two billion years later, when successful colonies of individual bacteria must have seemed to own the earth, it was the turn of single cells equipped with a nucleus. Bacteria were single living organisms too, but their DNA had not been collected in a nucleus. Single cells with a nucleus were a notch up. These life-forms were known technically as eukaryotic cells, which belong to a large group of organisms, the Protozoa. Back in the morning of life, such cells were some of the first truly independent organisms. Each of them could survive individually without symbiotic partnerships. Such simple single organisms are still with us today. The lively amoeba is a good example, and so is the wonderful paramecium.[1]

A single cell has a body frame (a cytoskeleton), inside which there is a nucleus (the command center that houses the cell's DNA) and a cytoplasm (where the transformation of fuel into energy takes place under the control of organelles such as mitochondria). Bodies are demarcated by skins, and the cell does have one, a boundary between its interior and the exterior world. It is called the cell membrane.

In many respects a single cell is a preview of what a single organism such as ours would come to be. One can see it as a sort of cartooned abstraction of what we are. The cytoskeleton is the scaffolding frame of the body proper, just as the bone skeleton is in all of us. The cytoplasm corresponds to the interior of the body proper with all its organs. The nucleus is the equivalent of the brain. The cell membrane is the equivalent of the skin. Some of these cells even have the equivalent of limbs, cilia, whose concerted movements allow them to swim.

The separate components of a eukaryotic cell came together by way of cooperation among simpler individual creatures, namely, bacteria that gave up their independent status to be a part of a convenient new aggregate. A certain kind of bacterium gave rise to mitochondria; another kind, such as spirochetes, helped with the cytoskeleton and with cilia, for those that liked to swim, and so forth.[2] The marvel is that each of our own multicellular organisms is put together according to this same basic strategy, by aggregating billions of cells so as to constitute tissues, pulling together different kinds of tissue so as to constitute organs, and connecting different organs so as to form systems. Examples of tissues include the epithelia of skin, mucosal linings and endocrine glands, the muscular tissue, the nervous or neural tissue, and the connective tissue that binds them all in place. Examples of organs are obvious, from hearts and guts to the brain. Examples of systems include the ensemble formed by the heart, blood, and blood vessels (the circulatory system), the immune system, and the nervous system. As a result of this cooperative arrangement, our organisms are highly differentiated combinations of trillions of cells of varied kinds, including, of course, neurons, the most distinctive constituents of the brain. More about neurons and brains in a minute.

The main difference between the cells found in multicellular (or metazoan) organisms and the cells of unicellular organisms is that while single cells must fend for themselves, the cells that constitute multicellular organisms live within highly diverse, complex societies. Many of the tasks that the cells of unicellular organisms must accomplish alone are, in multicellular organisms, assigned to specialized cell types. The general arrangement is comparable to the diverse assignment of functional roles that each individual cell embodies in its own structure. Multicellular organisms are made of multiple, cooperatively organized unicellular organisms, which first arose from the combination of even smaller individual organisms. The economy of a multicellular organism has many sectors, and the cells within those sectors cooperate. If this sounds familiar and makes you think of human societies, it is because it should. The resemblances are staggering.

The governance of a multicellular organism system is highly decentralized, although it does have leadership centers with advanced powers of analysis and decision, like the endocrine system and the brain. Still, with rare exceptions, all the cells in multicellular organisms, ours included, have the same components as those of a single one—membrane, cytoskeleton, cytoplasm, nucleus. (Red blood cells, whose brief, 120-day life is devoted to transporting hemoglobin, are the exception: they have no nucleus to speak of.) Moreover, all those cells have a comparable life cycle—birth, development, senescence, death—as does a big organism. The life of a single human organism is built of multitudes of simultaneous, well-articulated lives.

As simple as they were and are, single cells had what appeared to be a decisive, unshakable determination to stay alive for as long as the genes inside their microscopic nucleus commanded them to do so. The governance of their life included a stubborn insistence to remain, endure, and prevail until such time as some of the genes in the nucleus would suspend the will to live and allow the cell to die.

I know it is difficult to imagine that the notions of "desire" and "will" are applicable to a single lonely cell. How can attitudes and intentions that we associate with the conscious human mind, and that we intuit to result from the workings of big human brains, be present at such an elementary level? But there they are, by whatever name you may wish to call those features of the cell's behavior.[3]

Deprived of conscious knowledge, deprived of access to the byzantine devices of deliberation available in our brains, the single cell seems to have an attitude: it wants to live out its prescribed genetic allowance. Strange as it may seem, the want, and all that is necessary to implement it, *precedes* explicit knowledge and deliberation regarding life conditions, since the cell clearly has neither. The nucleus and the cytoplasm interact and carry out complex computations aimed at keeping the cell alive. They deal with the moment-to-moment problems posed by the living conditions and adapt the cell to the situation in a survivable manner. Depending on the environmental conditions, they rearrange the position and distribution of molecules in their interior, and they change the shape of subcomponents, such as microtubules, in an astounding display of precision. They respond under duress and under nice treatment too. Obviously, the cell components carrying out those adaptive adjustments were put into place and instructed by the cell's genetic material.

We commonly fall into the trap of regarding our big brains and complex conscious minds as the originators of the attitudes, intentions, and strategies behind our sophisticated life management. Why should we not? That is a reasonable and parsimonious way of conceiving the history of such processes, when we view it from the top of the pyramid and from present circumstances. The reality, however, is that the conscious mind has merely made the basic life-management know-how, well, *knowable*. As we shall see, the decisive contributions of the conscious mind to evolution come at a much higher level; they have to do with deliberative, offline decision-making and with cultural creations. I am definitely not minimizing the importance of that high level of life management. Indeed, one of the main ideas in this book is that the human conscious mind has taken evolution in a new course precisely by providing us with choices, by making relatively flexible sociocultural regulation possible beyond the complex social organization that social insects, for example, so spectacularly exhibit. Rather, I am reversing the narrative sequence of the traditional account of consciousness by having covert knowledge of life management *precede* the conscious experience of any such knowledge. I am also saying that the covert knowledge is quite sophisticated and should not be regarded as primitive. Its complexity is huge and its seeming intelligence remarkable.

I am not downgrading consciousness but am most certainly upgrading nonconscious life management and suggesting that it constitutes the blueprint for attitudes and intentions of conscious minds.

Every cell in our body has the kind of nonconscious attitude I have just described. Could it be that our very human conscious desire to live, our will to prevail, began as an aggregate of the inchoate wills of all the cells in our body, a collective voice set free in a song of affirmation?

The notion of a large collective of wills expressed through one single voice is not mere poetic fancy. It connects with the reality of our organisms where that single voice does exist in the form of the *self* in a conscious brain. But how does one transfer the brainless, mindless wills of single cells and their collectives to the self of conscious minds that originates in a brain? For that to happen, we need to introduce a radical, game-changing actor in our narrative: the nervous cell or neuron.

Neurons, as far as one can fathom, are unique cells, of a kind unlike any other in the body, unlike even other kinds of brain cells such as glial cells. What makes neurons so different and so special? After all, don't they too have a cell body, equipped with nucleus, cytoplasm, and membrane? Don't they rearrange molecules internally as other body cells do? Don't they too adapt to the environment? Yes, indeed, all the above is true. Neurons are, through and through, body cells, and yet they also are special.

To explain why neurons are special, we should consider a functional difference and a strategic difference. The essential functional difference has to do with the neuron's ability to produce electrochemical signals capable of changing the state of other cells. Neurons did not invent electrical signals. For example, unicellular organisms such as paramecia can also produce them and use them to govern their behavior. But neurons use their signals to influence other cells, namely, other neurons, endocrine cells (which secrete chemical molecules), and muscle fiber cells. Changing the state of other cells is the very source of the activity that constitutes and regulates behavior, to begin with, and that eventually also contributes to making a mind. Neurons are capable of this feat because they produce and propagate an electrical current along the tubelike section known as the axon. Sometimes the transmission goes over distances that can be appreciated by the naked eye, as when signals travel for many centimeters along the axons of neurons from our motor cortex to the brain stem, or from the spinal cord to the tip of a limb. When the electrical current arrives at the tip of the neuron, the synapse, it causes the release of a chemical molecule, a transmitter, which in turn acts on the subsequent cell in the chain. When the subsequent cell is a muscle fiber, movement ensues.[4]

There is no longer any mystery as to why neurons do this. Like other body cells, neurons have electrical charges on the inside and outside of their membranes. The charges are due to the concentration of ions such as sodium or potassium on either side of the wall. But neurons take advantage of creating large charge differences between inside and outside—the state of polarization. When this difference is drastically reduced, at a certain point in the cell, the membrane depolarizes locally, and the depolarization advances down the axon as if it were a wave. That wave is the electrical impulse. When neurons depolarize, we say they are "on," or "firing." In brief, neurons are like other cells, but they can send influential signals to other cells and thus modify what those other cells do.

The above functional difference is responsible for a major strategic difference: *neurons exist for the benefit of all the other cells in the body.* Neurons are not essential for the basic life process, as all those living creatures that have no neurons at all easily demonstrate. But in complicated creatures with many cells, neurons *assist* the multicellular body proper with the management of life. That is the purpose of neurons and the purpose of the brains they constitute. All the astonishing feats of brains that we so revere, from the marvels of creativity to the noble heights of spiritual-ity, appear to have come by way of that determined dedication to man-aging life within the bodies they inhabit.

Even in modest brains, made of networks of neurons arranged as ganglia, neurons assist other cells in the body. They do so by receiving signals from body cells and either promoting the release of chemical molecules (as they do with a hormone secreted by an endocrine cell that reaches body cells and changes their function) or by making movements happen (as when neurons excite muscle fibers and make them contract). In the elaborate brains of complex creatures, however, networks of neurons eventually come to mimic the structure of parts of the body to which they belong. They end up *representing* the state of the body, liter-ally mapping the body for which they work and constituting a sort of virtual surrogate of it, a neural double. Importantly, they remain con-nected to the body they mimic throughout life. As we shall see, mim-icking the body and remaining connected to it serve the managing function quite well.

In brief, neurons are *about* the body, and this "aboutness," this relentless pointing to the body, is the defining trait of neurons, neuron circuits, and brains. I believe this aboutness is the reason why the covert will to live of the cells in our body could ever have been translated into a minded, conscious will. The covert, cellular wills came to be mimicked by brain circuitry. Curiously, the fact that neurons and brains are about the body also suggests how the external world would get mapped in the brain and mind. As I will explain in Part II, when the brain maps the world external to the body, it does so thanks to the mediation of the body. When the body interacts with its environment, changes occur in the body's sensory organs, such as the eyes, ears, and skin; the brain maps those changes, and thus the world outside the body indirectly acquires some form of representation within the brain.

In closing this hymn to the particularity and glory of neurons, let me add a note on their origin and make them somewhat more modest. Evolutionarily, neurons probably arose from eukaryotic cells that commonly changed their shape and produced tubelike extensions of their body as they moved about, sensing the environment, incorporating food, going about the business of life. The pseudopodia of an amoeba give the gist of the process. The tubelike prolongations, which are created on the spot by internal rearrangements of microtubules, are dismantled once the cell has accomplished its business. But when such temporary prolongations became permanent, they became the tubelike components that make neurons so distinctive—the axons and the dendrites. A stable collection of cable work and antennas, ideal to emit and receive signals, was born.[5]

Why is this important? Because while the operation of neurons is quite distinctive and opened the way for complex behavior and mind, neurons maintained a close kinship to other body cells. Simply looking at neurons and at the brains they constitute as radically different cells without taking their origins into account risks separating the brain from the body further than is justifiable, given its genealogy and operation. I suspect that a good part of the puzzlement regarding how feeling states can emerge in the brain derives from overlooking the deep body-relatedness of the brain.

One other distinction must be made between neurons and other body cells. To the best of our knowledge, neurons do not reproduce—that is, they do not divide. Nor do they regenerate, or at least not to a significant extent. Practically all other cells in the body do, although the cells of the lenses in our eyes and the muscle fiber cells of the heart are exceptions. It would not be a good idea for such cells to divide. If cells in the lens were to undergo division, the transparency of the medium would likely be affected during the process. If cells in the heart were to divide (even only one sector at a time, a bit like the carefully planned remodeling of a house), the pumping action of the heart would be severely compromised, much as it is when a myocardial infarct disables a sector of the heart and unbalances its chambers' fine coordination. What about the brain? Although we lack a complete understanding of how neuron circuits maintain memories, division of neurons would probably disrupt the records of a lifetime of experience that are inscribed, by learning, in particular patterns of neurons firing in complex circuits. For the same reason, division would also disrupt the sophisticated know-how that is inscribed in circuits by our genome from the get-go and that tells the brain how to coordinate the operations of life. Division of neurons might spell the end of species-specific life regulation and would possibly not allow behavioral and mental individuality to develop, let alone become identity and personhood. The plausibility of this dire scenario is in the known consequences of damage to certain neuron circuits as caused by stroke or Alzheimer's disease.

The division of most other cells in our bodies is highly regimented, so as not to compromise the architecture of the varied organs and the overall architecture of the organism. There is a *Bauplan* that must be adhered to. Throughout the life span, a continuous *restoration* is going on rather than genuine remodeling. No, we do not knock down walls in our body house; nor do we build a new kitchen or add a guest wing. The restoration is very subtle, quite meticulous. For a good part of our lives, the substitution of cells is so perfectly achieved that even our appearance remains the same. But when one considers the effects of aging relative to the external appearance of our organism or to the operation of our internal system, one realizes that the substitutions become gradually less perfect. Things are not quite in the same place. The skin of the face ages, muscles sag, gravity intervenes, organs may not work quite so well. And that is when a good Beverly Hills plastic surgeon and efficient concierge medicine should enter the picture.

Staying Alive

What does it take for a living cell to stay alive? Quite simply, it takes good housekeeping and good external relations, which is to say good management of the myriad problems posed by living. Life, in a single cell as well as in large creatures with trillions of them, requires the transformation of suitable nutrients into energy, and that, in turn, calls for the ability to solve several problems: finding the energy products, placing them inside the body, converting them into the universal currency of energy known as ATP, disposing of the waste, and using the energy for whatever the body needs to continue this same routine of finding the right stuff, incorporating it, and so forth. Procuring nutrition, consuming and digesting it, and allowing it to power a body—those are the issues for the humble cell.

The mechanics of life management are crucial because of its difficulty. Life is a precarious state, made possible only when a large number of conditions are met simultaneously within the body's interior. For example, in organisms such as ours, the amounts of oxygen and CO_2 can vary only within a narrow range, as can the acidity of the bath in which chemical molecules of every sort travel from cell to cell (the pH). The same applies to temperature, whose variations we are keenly aware of when we have a fever or, more commonly, when we complain of the weather's being too hot or too cold; it also applies to the amount of fundamental nutrients in circulation—sugars, fats, proteins. We feel discomfort when the variations depart from the nice and narrow range, and we feel quite agitated if we go for a very long time without doing something about the situation. These mental states and behaviors are signs that the ironclad rules of life regulation are being disobeyed; they are prompts from the netherlands of nonconscious processing toward minded and conscious life, requesting us to find a reasonable solution for a situation that can no longer be managed by automatic, nonconscious devices.

When one measures each of those parameters and attributes numbers to them, one discovers that the range within which they normally vary is extremely small. In other words, life requires that the body maintain a collection of parameter *ranges* at all costs for literally dozens of components in its dynamic interior. All the management operations to which I alluded earlier—procuring energy sources, incorporating and transforming energy products, and so forth—aim at maintaining the chemical parameters of a body's interior (its internal milieu) within the magic range compatible with life. The magic range is known as *homeostatic,* and the process of achieving this balanced state is called *homeostasis.* These not-so-elegant words were coined in the twentieth century by the physiologist Walter Cannon. Cannon expanded on the discoveries of the nineteenth-century French biologist Claude Bernard, who had coined the nicer term *milieu intérieur* (internal milieu), the chemical soup within which the struggle for life goes on uninterrupted but hidden from view. Unfortunately, although the essentials of life regulation (the process of homeostasis) have been known for more than a century and are applied daily in general biology and medicine, their deeper significance in terms of neurobiology and psychology has not been appreciated.[6]

The Origins of Homeostasis

How was homeostasis ever planted in whole organisms? How did single cells acquire their life-regulation design? To approach such a question, one must engage in a problematic form of reverse engineering that is never easy because we have spent most of our scientific history thinking from the perspective of whole organisms rather than from the perspective of the molecules and genes with which organisms began.

The fact that homeostasis began unknowingly, at the level of organisms without consciousness, mind, or brain, raises the question of where and how the homeostatic intention was planted in the history of life. That question takes us down from single cells to genes and from there to simple molecules, simpler even than DNA and RNA. The homeostatic intention may arise from those simple levels and may even be related to the basic physical processes that govern the interaction of molecules— for example, the forces with which two molecules attract or reject each other, or combine constructively or destructively. Molecules repulse or attract; they assemble and participate explosively, or they refuse to do so.

As far as organisms are concerned, gene networks resulting from natural selection were evidently responsible for endowing them with homeostatic capability. What kind of knowledge did (and do) gene networks possess in order to be able to pass on such wise instructions to the organisms they launched? Where is the origin of value—its "primitive"—when we go below the level of tissues and cells to the level of genes? Perhaps what is needed is a specific ordering of genetic information. At the level of gene networks the primitive of value would consist of an ordering of gene expression that would result in the construction of "homeostatically competent" organisms.

But deeper answers must be sought at even simpler levels. There are important debates regarding how the process of natural selection has operated to produce the human brains we currently enjoy. Has natural selection operated at the level of genes, or whole organisms, or groups of individuals, or all of the above? But from the gene perspective, and in order for genes to survive over generations, gene networks had to construct perishable and yet successful organisms that served as vehicles. And in order for organisms to behave in such a successful manner, genes must have guided their assembly with some critical instructions.

A good part of those instructions must have consisted of constructing devices capable of conducting efficient life regulation. The newly assembled devices handled the distribution of rewards, the application of punishments, and the prediction of situations that an organism would face. In brief, gene instructions led to the construction of devices capable of executing what, in complex organisms like us, came to flourish as emotions, in the broad sense of the term. The early sketch of these devices was first present in organisms without brain, mind, or consciousness—the single cells we discussed earlier; however, the regulating devices attained the greatest complexity in organisms that have all three: brain, mind, and consciousness.[7]

Is homeostasis enough to guarantee survival? Not really, because attempting to correct homeostatic imbalances after they begin is inefficient and risky. Evolution took care of this problem by introducing devices that allow organisms to anticipate imbalances and that motivate them to explore environments likely to offer solutions.

Cells, Multicellular Organisms, and Engineered Machines

Cells and multicellular organisms share several features with engineered machines. The activity of either living organisms or engineered machines achieves a goal; there are component processes to the activity; the processes are executed by distinct anatomical parts that perform subtasks; and so forth. The resemblance is quite suggestive and is behind the two-way metaphors with which we describe both living things and machines. We talk of the heart as a pump, we describe blood circulation as plumbing, we refer to the action of limbs as levers, and so forth. Likewise, when we consider an indispensable operation in a complex machine, we call it the "heart" of the machine, and we refer to the control devices of the same machine as its "brain." Machines that operate unpredictably are called "temperamental." This mode of thinking, which by and large is quite illuminating, is also responsible for the less-than-helpful idea that the brain is a digital computer and the mind something like the software that one may run in it. But the real problem of these metaphors comes from their neglect of the fundamentally different statuses of the *material components* of living organisms and engineered machines. Compare a modern marvel of aircraft design—the Boeing 777—with any example of living organism, small or large. A number of similarities can easily be identified—command centers in the form of cockpit computers; feedforward information channels to those computers, regulating feedback channels to the peripheries; a metabolism of sorts present in the fact that engines feed on fuel and transform energy; and so forth. And yet a fundamental difference persists: any living organism is naturally equipped with global homeostatic rules and devices; in case they malfunction, the living organism's body perishes; even more important, *every* component of the living organism's body (by which I mean every cell) is, in itself, a living organism, naturally equipped with its own homeostatic rules and devices, subject to the same risk of perishability in case of malfunction. The structure of the admirable 777 has nothing comparable whatsoever, from its metal-alloy fuselage to the materials that make up its miles of wiring and hydraulic tubing. The high-level "homeostatics" of the 777, shared by its bank of intelligent on-board computers and the two pilots needed to fly the aircraft, aim at preserving its entire, one-piece structure, not its micro and macro physical subcomponents.

Biological Value

As I see it, the most essential possession of any living being, at any time, is the balanced range of body chemistries compatible with healthy life. It applies equally to an amoeba and to a human being. All else flows from it. Its significance cannot be overemphasized.

The notion of biological value is ubiquitous in modern thinking about brain and mind. We all have an idea, or perhaps several ideas, of what the word *value* means, but what about *biological value*? Let us consider some other questions: Why do we take virtually everything that surrounds us—food, houses, gold, jewelry, paintings, stocks, services, even other people—and assign a value to it? Why does everyone spend so much time calculating gains and losses related to those items? Why do items carry a price tag? Why this incessant valuation? And what are the yardsticks against which value is measured? At first glance these questions might seem to have no place in a conversation about brain, mind, and consciousness. But in fact they do, and, as we shall see, the notion of value is central to our understanding of brain evolution, brain development, and actual, moment-to-moment brain activity.

Of the questions posed above, only the question of why items carry a price tag has a fairly straightforward answer. Indispensable items and items that are hard to obtain, given the high demand for them or their relative rarity, carry a higher cost. But why do they need a price? Well, there is not possibly enough of everything for everyone to have some; pricing is a means to govern the very real mismatch between what is available and the demand for it. Pricing introduces restraint and creates some sort of order in the access to items. But why is there not enough of everything for everyone? One reason has to do with the uneven distribution of needs. Certain items are very much needed, others less so, some not at all. Only when we introduce the notion of need do we come, finally, to the crux of biological value: the matter of a living individual struggling to maintain life and the imperative needs that arise in the struggle. Why we assign value in the first place, however, or the choice of yardstick we use in the assignment, requires an acknowledgment of the problem of maintaining life and of its requisite needs. As far as humans are concerned, maintaining life is only part of a larger problem, but let us stay with survival to begin with.

To date, neuroscience has dealt with this set of questions by taking a curious shortcut. It has identified several chemical molecules that are related, in one way or another, to states of reward or punishment and thus, by extension, are associated with value. Some of the best-known molecules will sound familiar to many readers: dopamine, norepinephrine, serotonin, cortisol, oxytocin, vasopressin. Neuroscience has also identified a number of brain nuclei that manufacture such molecules and deliver them to other parts of the brain and the body. (Brain nuclei are collections of neurons located below the cerebral cortex in the brain stem, hypothalamus, and basal forebrain; they should not be confused with the nuclei inside eukaryotic cells, which are simple sacs where most of the cell's DNA is housed.)[8]

The complicated neural mechanics of "value" molecules is an important topic that many committed neuroscience researchers are attempting to unravel. What prompts the nuclei to release those molecules? Where in the brain and body are they released precisely? What does their release accomplish? Somehow discussions about the fascinating new facts come up short when one turns to the central question: *Where is the engine for the value systems? What is the biological primitive of value?* In other words, where is the impetus for this byzantine machinery? Why did it even begin? Why did it turn out to be this way?

Without a doubt, the popular molecules and their nuclei of origin are important parts of the machinery of value. But they are not the answer to the questions posed above. I see value as indelibly tied to need, and need as tied to life. The valuations we establish in everyday social and cultural activities have a direct or indirect connection with homeostasis. That connection explains why human brain circuitry has been so extravagantly dedicated to the prediction and detection of gains and losses, not to mention the promotion of gains and the fear of losses. It explains, in other words, the human obsession with assignation of value.

Value relates directly or indirectly to survival. In the case of humans in particular, value also relates to the *quality* of that survival in the form of *well-being*. The notion of survival—and, by extension, the notion of biological value—can be applied to varied biological entities, from molecules and genes to whole organisms. I shall consider the whole organism perspective first.

Biological Value in Whole Organisms

Crudely stated, the paramount value for whole organisms consists of healthy survival to an age compatible with reproductive success. Natural selection has perfected the machinery of homeostasis to permit precisely that. Accordingly, the physiological state of a living organism's tissues, within an optimal homeostatic range, is the deepest origin of biological value and valuations. The statement applies equally to multicellular organisms and to those whose living "tissue" is confined to one cell.

The ideal homeostatic range is not absolute—it varies according to the context in which an organism is placed. But toward the extremes of the homeostatic range, the viability of living tissue declines and the risk of disease and death increases; within a certain sector of the range, however, living tissues flourish and their function becomes more efficient and economic. Operating near the extremes of the range, if only for brief periods of time, is actually an important advantage in unfavorable life conditions, but nonetheless life states operating close to the efficient range are preferable. It is reasonable to conclude that the primitive of organism value is inscribed in the configurations of physiological parameters. Biological value moves up or down a scale relative to the life-effectiveness of the physical state. In a way, biological value is a surrogate of physiological efficiency.

My hypothesis is that objects and processes we confront in our daily lives acquire their assigned value by reference to this primitive of naturally selected organism value. The values that humans attribute to objects and activities would bear some relation, no matter how indirect or remote, to the two following conditions: first, the general maintenance of living tissue within the homeostatic range suitable to its current context; second, the particular regulation required for the process to operate within the sector of the homeostatic range associated with well-being relative to the current context.

For whole organisms, then, the primitive of value is *the physiological state of living tissue within a survivable, homeostatic range.* The continuous representation of chemical parameters within the brain allows nonconscious brain devices to *detect and measure* departures from the homeostatic range and thus act as sensors for the degree of internal need. In turn, the measured departure from homeostatic range allows yet other brain devices to command corrective actions and even to promote *incentive* or *disincentive* for corrections, depending on the urgency of response. A simple record of such proceedings is the basis of the *prediction* of future conditions.

In brains capable of representing internal states in the form of maps, and potentially having minds and consciousness, the parameters associated with a homeostatic range correspond, at conscious levels of processing, to the *experiences* of pain and pleasure. Subsequently, in brains capable of language, those experiences can be assigned specific linguistic labels and called by their *names*—pleasure, well-being, discomfort, pain.

If you turn to a standard dictionary and look up the word *value,* you will find something like the following: "relative worth (monetary, material, or otherwise); merit; importance; medium of exchange; amount of something that can be exchanged for something else; the quality of a thing which renders it desirable or useful; utility; cost; price." As you can see, biological value is the root of all those meanings.

The Success of Our Early Forerunners

What made organism-vehicles so brilliantly successful? What opened the way for complex creatures such as ours? One important ingredient for our arrival appears to be something plants do not have but that we and some other animals do: *movement*. Plants can have tropisms; some can turn to or away from the sun and the shadows; and some, like the carnivorous Venus flytrap, can even catch distracted insects; but no plant can uproot itself and go seek a better environment in another part of the garden. The gardener must do that for it. The tragedy of plants, though they do not know it, is that their corseted cells could never change their shape enough to become neurons. Plants have no neurons and, in the absence of neurons, never a mind.

Independent organisms without brains also developed another important ingredient: the ability to *sense* changes in physiological condition, inside their own perimeter and in their surround. Even bacteria respond to sunlight as well as to countless molecules; bacteria in a petri dish will respond to the drop of a toxic substance by clumping together and recoiling from the threat. Eukaryotic cells also sensed the equivalent of touch and vibration. The changes sensed either in the interior or in the surrounding environment could lead to movement from one place to another. But in order to respond to a situation in an effective manner, the brain equivalent of single cells also has to harbor a *response policy,* a set of extremely simple rules according to which it makes a "decision to move" when certain conditions are met.

In brief, the minimal features that such simple organisms had to have so that they could succeed and let their genes travel on to the next generation were *sensing* of the organism's interior and exterior, a *response policy,* and *movement*. Brains evolved as devices that could improve the business of sensing, deciding, and moving and run it in more and more effective and differentiated manner.

Movement was eventually refined, thanks to the development of striated muscle, the kind of muscle we use today to walk and speak. As we shall see in Chapter 3, the sensing of the organism's interior, what we now call *interoception,* expanded to detect a large number of parameters (e.g., pH, temperature, presence or absence of numerous chemical molecules, tension of smooth muscle fibers). As for sensing the exterior, it came to include smell, taste, touch and vibration, hearing, and seeing, the ensemble of which we designate as *exteroception.*

For movement and sensing to work to the best advantage, the response policy must be something akin to an encompassing business plan outlining implicitly the conditions that inform the policy. This is precisely what the *homeostatic design* that we find in creatures of all levels of complexity consists of: a collection of operation guidelines that must be followed for the organism to achieve its goals. The essence of the guidelines is quite simple: if this is present, then do that.

When one surveys the spectacle of evolution, one is astounded by its many accomplishments. Consider, for example, the successful development of eyes, not only eyes that resemble ours but other varieties of eyes that do their job using slightly different means. No less astonishing is the marvel of echolocation, which allows bats and barn owls to hunt in complete darkness guided by exquisite sound localization in three-dimensional space. The evolution of a response policy capable of leading organisms to a homeostatic state is no less spectacular.

The rhyme and reason behind the existence of a response policy is the achievement of a homeostatic goal. But as I hinted earlier, even with a clear-cut goal, something else is needed for a response policy to be executed effectively. For a certain action to be achieved expeditiously and correctly, there must be an *incentive* so that, in certain circumstances, certain kinds of responses can be favored over others. Why? Because some circumstances of the living tissue may be so dire that they require urgent and decisive correction, and a literally breathless correction must be rapidly deployed. Likewise, some opportunities may be so conducive to the betterment of the living tissue that responses endorsing those opportunities must be selected and engaged rapidly. This is where we find the machinations behind what we have come to know, from our human perspective, as reward and punishment, the lead players in the dance of motivated exploration. Note that none of these operations requires a mind, let alone a conscious mind. There is no formal "subject" inside or outside an organism behaving as a "rewarder" or as a "punisher." Yet "rewards" and "punishments" are administered based on the design of response policy systems. The entire operation is as blind and "subject-less" as gene networks themselves are. Absence of mind and of self is perfectly compatible with spontaneous and implicit "intention" and "purpose." The basic "intention" of the design is to maintain structure and state, but a larger "purpose" can be construed from such multiple intentions: to survive.

What I am suggesting, then, is that *incentive* mechanisms are necessary to achieve successful guidance of behavior, which amounts to a successful, economic execution of the cell's business plan. I am also suggesting that the incentive mechanisms and the guidance did not arise from conscious determination and deliberation. There was no explicit knowledge and no deliberative self.

The guidance of incentive mechanisms has been made gradually more known to minded and conscious organisms such as ours. The conscious mind simply reveals what has long existed as an evolutionary mechanism of life regulation. But the conscious mind did not create the mechanism. The real story stands our intuition on its head. The actual historical sequence is reversed.

Developing Incentives

How did incentives develop? Incentives began in very simple organisms but are very evident in organisms whose brains are capable of measuring the *degree of* need for a certain correction. For the measurement to occur, the brain required a representation of (1) the *current* state of the living tissue, (2) the *desirable* state of the living tissue corresponding to the homeostatic goal, and (3) a simple comparison. Some kind of internal scale was developed for this purpose, signifying how far the goal was relative to the current state, while chemical molecules whose presence sped up certain responses were adopted to facilitate the correction. We are still sensing our organism states in terms of such a scale, something we do quite unconsciously, although the consequences of the measurement are made quite conscious when we feel hungry, very hungry, or not hungry at all.

What we have come to perceive as feelings of pain or pleasure, or as punishments or rewards, corresponds directly to integrated states of living tissue within an organism, as they succeed one another in the natural business of life management. The brain mapping of states in which the parameters of tissues depart significantly from the homeostatic range in a direction *not* conducive to survival is experienced with a quality we eventually called pain and punishment. Likewise, when tissues operate in the best part of the homeostatic range, the brain mapping of the related states is experienced with a quality we eventually named pleasure and reward.

The agents involved in orchestrating these tissue states are known as hormones and neuromodulators and were already very much present in simple organisms with only one cell. We know how these molecules operate. For example, in organisms with a brain, when a given tissue is risking its health due to a dangerously low level of nutrients, the brain detects the change and grades the need and the urgency with which the change must be corrected. This happens nonconsciously, but in brains with minds and consciousness, the state related to this information can become conscious. If and when it does, the subject experiences a negative feeling that may range from discomfort to pain. With or without consciousness of the process, a corrective chain of responses is engaged, in chemical and neural terms, helped by molecules that speed up the process. In the case of conscious brains, however, the consequence of the molecular process is not merely a correction of the imbalance: it is also a reduction of a negative experience such as pain and an experience of pleasure/reward. The latter comes, in part, from the life-conducive state the tissue may have now achieved. Eventually, the mere action of the incentive molecules is likely to place the organism in the functional configuration associated with pleasurable states.

The appearance of brain structures capable of detecting the likely delivery of "goods" or "threats" to the organism was also important. Specifically, beyond sensing the goods or the threats in and of themselves, brains began to use cues to *predict* the delivery. They would signal the coming of goods with the release of a molecule, such as dopamine or oxytocin; or the coming of threats with cortisol-releasing hormone or prolactin. The release would in turn optimize the behavior required to obtain or avoid the delivery of the stimulus. Likewise, they would use molecules to signal a miscue (a prediction error) and behave accordingly; they would differentiate between the coming of an expected item and an unexpected one by degrees of neuron firing and the corresponding degree of release of a molecule (e.g., dopamine). Brains also became capable of using the pattern of stimuli—for example, the *repetition* or *alternation* of stimuli—to predict what might be happening next. When two stimuli happened close to each other, that spelled the possibility that a third stimulus might be coming.

What did all this machinery achieve? First, a more or less urgent response depending on the circumstances—in other words, a *differential* response. Second, it achieved responses optimized by prediction.

Homeostatic design and its associated incentive and prediction devices protected the integrity of the living tissue inside an organism. Curiously, much the same machinery was co-opted to ensure that the organism would engage in reproductive behaviors that favored the passing of genes. Sexual attraction, sexual desire, and mating rituals are examples. On the surface, the behaviors associated with life regulation and with reproduction became separate, but the deeper goal was the same, and it is thus not surprising that the mechanics are shared.

As organisms evolved, the programs underlying homeostasis became more complex, in terms of the conditions that prompted their engagement and the range of results. Those more complex programs gradually became what we now know as drives, motivations, and emotions (see Chapter 5).

In brief, homeostasis needs help from drives and motivations, which complex brains provide abundantly, deployed with the help of anticipation and prediction and played out in the exploration of environments. Humans certainly have the most advanced motivational system, complete with endless curiosity, a keen scouting drive, and sophisticated warning systems regarding future needs, all meant to keep us on the good side of the railroad tracks.

Connecting Homeostasis, Value, and Consciousness

What we have come to designate as valuable, in terms of goods or actions, is directly or indirectly related to the possibility of maintaining a homeostatic range in the interior of living organisms. Moreover, we know that certain sectors and configurations within the homeostatic range are associated with optimal life regulation, while others are less efficient, and others still are closer to the danger zone. The danger zone is that within which disease and death can set in. It stands to reason that goods and actions that, in one way or another, will ultimately induce optimal life regulation will be regarded as most valuable.[9]

We already know how humans diagnose the optimal sector of the homeostatic range, without any need to have one's blood chemistries measured in a medical lab. The diagnosis requires no special expertise but merely the fundamental process of consciousness: *optimal ranges express themselves in the conscious mind as pleasurable feelings; dangerous ranges, as not-so-pleasant or even painful feelings.*

Can one imagine a more transparent detection system? Optimal workings of an organism, which result in efficient, harmonious states of life, constitute the very substrate of our primordial feelings of well-being and pleasure. They are the foundation of the state that, in quite elaborate settings, we call happiness. On the contrary, disorganized, inefficient, inharmonious life states, the harbingers of disease and system failure, constitute the substrate of negative feelings, of which, as Tolstoy observed so accurately, there are far more varieties than of the positive kind—an infinite assortment of pains and suffering, not to mention disgust, fears, anger, sadness, shame, guilt, and contempt.

As we shall see, the defining aspect of our emotional feelings is the conscious readout of our body states as modified by emotions; that is why feelings can serve as barometers of life management. This is also why, not surprisingly, feelings have been influencing societies and cultures and all their workings and artifacts ever since they became known to human beings. But long before the dawn of consciousness and the emergence of conscious feelings, in fact even before the dawn of minds as such, the configuration of chemical parameters was already influencing individual behaviors in simple creatures without brains to represent those parameters. This is quite sensible: unminded organisms had to rely on chemical parameters to guide the actions required to maintain their lives. This "blind" guidance encompassed considerably elaborate behaviors. The growth of different kinds of bacteria in a colony is guided by such parameters and can even be described in social terms: colonies of bacteria routinely practice "quorum sensing" within their group and literally engage in warfare in order to hold on to territory and resources. They do that even inside our own bodies as they fight for real estate privileges in our throats or in our guts. But as soon as very simple nervous systems came on the scene, social behaviors were even more apparent. Consider the nematode, a polite name for a scientifically fetching kind of worm whose social behaviors are quite sophisticated.

The brain of a nematode, such as *C. elegans,* has a mere 302 neurons organized in a chain of ganglia—nothing to be very proud of. Like any other living creature, nematodes need to feed themselves to survive. Depending on the scarcity or abundance of food and on environmental threats, they can come to the trough, as it were, more or less gregariously. They feed alone if food is available and the environment is quiet; but if food is scarce or if they detect a threat in the environment (for instance, a certain kind of odor), they will come in groups. Needless to say, they do not really know what they are doing, let alone why. But they do what they do because their exceedingly simple brains, without any mind to speak of and even less proper consciousness, use signals from the environment to engage one kind of behavior or the other.

Now imagine that I had described the situation of *C. elegans* in the abstract, outlining the conditions and the behaviors but withholding the fact that they were worms, and now imagine that I had asked you to think as a sociologist and comment on the situation. I suspect you would have detected evidence of interindividual cooperation, and you might even have diagnosed altruistic concerns. You might even have thought that I was speaking of complex creatures, perhaps early humans. The first time I read Cornelia Bargmann's description of these findings, I thought of trade unions and of safety in numbers.[10] And yet *C. elegans* is just a worm.

Another implication of the fact that ideal homeostatic states are the most valuable possession of a living organism is that the fundamental advantage of consciousness, at any level of the phenomenon, derives from improving life regulation in ever more complex environments.[11]

Survival in new ecological niches was helped by brains complex enough to create minds, a development that, as I explain in Part II, is based on the construction of neural maps and images. Once minds emerged, even if they were not yet imbued with full-scale consciousness, automated life regulation was optimized. Brains that produced images had available more details of the conditions inside and outside the organisms and thus could generate more differentiated and effective responses than unminded brains. However, when the minds of nonhuman species could become conscious minds, automated regulation gained a powerful ally, a means to focus the travails of survival on the budding self that now stood for the struggling organism. In humans, of course, as consciousness coevolved with memory and reason to permit offline planning and deliberative thinking, that ally has become even more powerful.

Amazingly, self-concerned life regulation always coexists with the machinery of automated life regulation that any conscious creature inherited from its evolutionary past. This is very true of humans. Most of our own regulatory activity goes on unconsciously, and a good thing too. You would not want to manage your endocrine system or your immunity *consciously* because you would have no way of controlling chaotic oscillations rapidly enough. At best, it would be the equivalent of flying a modern jet plane by hand—which is not a trivial undertaking and does require one to master all the contingencies and all the maneuvers needed to prevent a stall. At worst, it would be like investing the Social Security Trust Fund in the stock market. You would not even want to have absolute control over something as simple as your breathing—you might decide to swim the English Channel underwater, holding your breath, and risk dying in the process. Fortunately, our automated homeostatic devices will never allow such foolery.

Consciousness has improved adaptability and allowed the beneficiaries to create novel solutions to the problems of life and survival, in virtually any conceivable environment, anywhere on earth, up in the air and in outer space, under the water, in deserts and on mountains. We have evolved to *adapt* to a large number of niches and are able to *learn to adapt* to an even greater number. We never sprouted wings or gills, but we invented machines that have wings or can rocket us into the stratosphere, that sail the oceans or travel twenty thousand leagues under those oceans. We have invented the material conditions to live anywhere we wish. The amoeba cannot; nor can the worm, the fish, the frog, the bird, the squirrel, the cat, the dog, or even our very smart cousin, the chimpanzee.

When human brains began concocting conscious human minds, the game changed radically. We moved from simple regulation, focused on the survival of the organism, to progressively more deliberated regulation, based on a mind equipped with identity and personhood and now actively seeking not mere survival but certain ranges of well-being. Quite a leap, albeit assembled, so far as we can see, on biological continuities.

If brains prevailed in evolution because they offered a larger compass of life regulation, the brain systems that led to conscious minds prevailed because they offered the widest possibilities of adaptation and survival with the sort of regulation capable of maintaining and expanding well-being.

In brief, single-cell organisms with a nucleus have an unminded and unconscious will to live and manage life suitably enough, for as long as certain genes allow them. Brains expanded the possibilities of life management even when they did not produce minds, let alone conscious minds. For that reason they too prevailed. By the time minds and consciousness were added to the mix, the possibilities of regulation expanded even more and made way for the kind of management that occurs not just within one organism but across many organisms, in societies. Consciousness enabled humans to repeat the leitmotif of life regulation by means of a collection of cultural instruments—economic exchange, religious beliefs, social conventions and ethical rules, laws, arts, science, technology. Still, the survival intention of the eukaryotic cell and the survival intention implicit in human consciousness are one and the same.

Behind the imperfect but admirable edifice that cultures and civilizations have constructed for us, life regulation remains the basic issue we face. Just as important, the motivation behind most achievements in human cultures and civilizations rests with that precise issue and with the need to manage the behaviors of humans engaged in addressing that issue. Life regulation is at the root of a lot that needs explaining in biology in general and in humanity in particular: the existence of brains; the existence of pain, pleasure, emotions, and feelings; social behaviors;

religions; economies and their markets and financial institutions; moral behaviors; laws and justice; politics; art, technology, and science— a very modest list, as the reader can see.

Life and the conditions that are integral to it—the irrepressible mandate to survive and the complicated business of managing survival in an organism, with one cell or with trillions—were the root cause of the emergence and evolution of brains, the most elaborate management devices assembled by evolution, as well as the root cause of everything that followed from the development of ever more elaborate brains, inside ever more elaborate bodies, living in ever more complex environments.

When one looks at most any aspect of brain functions through the filter of this idea—that a brain exists for managing life inside a body— the oddities and mysteries of some of the traditional categories of psychology (emotion, perception, memory, language, intelligence, *and* consciousness) become less odd and far less mysterious. In fact, they develop a transparent reasonableness, an inevitable and endearing logic. How could we be any different, those functions seem to be asking, given the job that needs to be done?

PART II

What's in a Brain That a Mind Can Be?

3

Making Maps and Making Images

Maps and Images

While the management of life is unquestionably the primary function of human brains, it is hardly their most distinctive feature. As we have seen, life can be managed without a nervous system, let alone a full-fledged brain. Modest unicellular organisms do pretty well at housekeeping.

The distinctive feature of brains such as the one we own is their uncanny ability to create maps. Mapping is essential for sophisticated management, mapping and life management going hand in hand. When the brain makes maps, it *informs* itself. The information contained in the maps can be used nonconsciously to guide motor behavior efficaciously, a most desirable consequence considering that survival depends on taking the right action. But when brains make maps, they are also creating images, the main currency of our minds. Ultimately consciousness allows us to experience maps as images, to manipulate those images, and to apply reasoning to them.

Maps are constructed when we interact with objects, such as a person, a machine, a place, from the outside of the brain toward its interior. I cannot emphasize the word *interaction* enough. It reminds us that making maps, which is essential for improving actions as noted above, often occurs in a setting of action to begin with. Action and maps, movements and mind, are part of an unending cycle, an idea suggestively captured by Rodolfo Llinás when he attributes the birth of the mind to the brain's control of organized movement.[1]

Maps are also constructed when we recall objects from the inside of our brain's memory banks. The construction of maps never stops even in our sleep, as dreams demonstrate. The human brain maps whatever object sits outside it, whatever action occurs outside it, and all the relationships that objects and actions assume in time and space, relative to each other and to the mother ship known as the organism, sole proprietor of our body, brain, and mind. The human brain is a born cartographer, and the cartography began with the mapping of the body inside which the brain sits.

The human brain is a mimic of the irrepressible variety. Whatever sits outside the brain—the body proper, of course, from the skin to the entrails, as well as the world around, man, woman, and child, cats and dogs and places, hot weather and cold, smooth textures and rough, loud sounds and soft, sweet honey and salty fish—is mimicked inside the brain's networks. In other words, the brain has the ability to represent aspects of the structure of nonbrain things and events, which includes the actions carried out by our organism and its components, like limbs, parts of the phonatory apparatus, and so forth. How the mapping happens exactly is easier said than done. It is not a mere copy, a passive transfer from the outside of the brain toward its inside. The assembly conjured by the senses involves an active contribution offered from inside the brain, available from early in development, the idea that the brain is a blank slate having long since lost favor.[2] The assembly often occurs in the setting of movement, as noted earlier.

A brief note on terminology: I used to be strict about using the term *image* only as a synonym of mental pattern or mental image, and the term *neural pattern* or *map* to refer to a pattern of activity *in the brain* as distinct from the mind. My intent was to recognize that the mind, which I see as inhering in the activity of brain tissue, deserves its own description because of the private nature of its experience, and because that private experience is precisely the phenomenon we wish to explain; as for describing neural events with their proper vocabulary, it was part of the effort to understand the role of those events in the mind process. By keeping separate levels of description, I was not suggesting at all that there are separate substances, one mental and the other biological. I am not a substance dualist as Descartes was, or tried to make us believe he was, by saying that the body had physical extension but the mind did not, as the two are made of different substances. I was simply indulging in *aspect* dualism and discussing the way things appear, on their experiential surface. But, of course, so did my friend Spinoza, the standard-bearer for monism, the very opposite of dualism.

But why complicate matters, for myself and for the reader, by using separate terms to refer to two things that I believe to be equivalent? Throughout this book, I use the terms *image, map,* and *neural pattern* almost interchangeably. On occasion I also blur the line between mind and brain, deliberately, to underscore the fact that the distinction, while valid, can block the view of what we are trying to explain.

Cutting Below the Surface

Imagine holding a brain in your hand and looking at the surface of the cerebral cortex. Now imagine taking a sharp knife and making cuts *parallel* to the surface, at a depth of two or three millimeters, and extracting a thin filet of brain. After fixing and staining the neurons with an appropriate chemical, you can lay your preparation down on a thin glass slide and look at it under the microscope. You will discover, in each cortical layer that you inspect, a sheathlike structure that essentially resembles a two-dimensional square grid. The main elements in the grid are neurons, displayed horizontally. You can imagine something like the plan of Manhattan, but you must leave Broadway out because there are no major oblique lines in the cortical grids. The arrangement, you immediately realize, is ideal for overt topographical representation of objects and actions.

Looking at a patch of cerebral cortex, it is easy to see why the most detailed maps the brain makes arise here, although other parts of the brain can also make them, albeit with a lower resolution. One of the cortical layers, the fourth, is probably responsible for a large part of the detailed maps. Contemplating a patch of cerebral cortex, one also realizes why the idea of brain maps is not a far-fetched metaphor. One can sketch patterns onto such a grid, and when one squints a little and lets the imagination roam free, one can picture the sort of parchment paper that Henry the Navigator probably pored over when he was planning the voyages of his captains. One big difference, of course, is that the lines in a brain map are not drawn with quill or pencil; they are, rather, the result of the momentary activity of some neurons and of the inactivity of others. When certain neurons are "on," in a certain spatial distribution, a line is "drawn," straight or curved, thick or thin, a pattern distinct from the background created by the neurons that are "off." Another big difference: the lead map-making horizontal layer is stacked between other layers above and below; each main element of the layer is also part of a vertical array of elements, namely, a column. Each column contains hundreds of neurons. Columns provide inputs to the cerebral cortex (the inputs come from elsewhere in the brain, from peripheral sensory probes such as the eyes, and from the body). Columns also provide outputs toward the same sources and carry out varied integrations and modulations of the signals being processed at each locality.

Brain maps are not static like those of classical cartography. Brain maps are mercurial, changing from moment to moment to reflect the changes that are happening in the neurons that feed them, which in turn reflect changes in the interior of our body and in the world around us. The changes in brain maps also reflect the fact that we ourselves are in constant motion. We come close to objects or move away from them; we can touch them and then not; we can taste a wine, but then the taste goes away; we hear music, but then it comes to an end; our own body changes with different emotions, and different feelings ensue. The entire environment offered to the brain is perpetually modified, spontaneously or under the control of our activities. The corresponding brain maps change accordingly.

A current analogy to what goes on in the brain relative to a visual map can be found in the sort of picture you see in electronic billboards, in which the pattern is drawn by active or inactive light elements (light-bulbs or light-emitting diodes). The analogy to electronic maps is all the more apt because the content depicted in them can rapidly change merely by changing the distribution of active versus inactive elements. Each distribution of activities constitutes a pattern in time. Different distributions of activity within the same patch of visual cortex can depict a cross, or a square, or a face, in succession or even in superposition. The maps can be rapidly drawn, redrawn, and overdrawn, at the speed of lightning.

The same kind of "drawing" also happens in an elaborate outpost of the brain called the retina. It too has a square grid ready to inscribe maps. When the light particles known as photons strike the retina in the particular distribution that corresponds to a specific pattern, the neurons activated by the pattern—say, a circle or a cross—constitute a transient neural map. Additional maps, based on the original retinal map, will be formed at subsequent levels of the nervous system. This is because the activity at each point in the retinal map is signaled forward along a chain, culminating in the primary visual cortices while preserving the geometrical relationships they hold at the retina, a property known as retinotopy.

Although the cerebral cortices excel at the creation of detailed maps, some structures below the cerebral cortex are able to create coarse maps. Examples include the geniculate bodies, the colliculi, the nucleus tractus solitarius, and the parabrachial nucleus. The geniculate bodies are dedicated, respectively, to visual and auditory processes. They too have a layered structure ideal for topographical representations. The superior colliculus is an important provider of visual maps and even has the ability to relate those visual maps to auditory and body-based maps. The inferior colliculus is dedicated to auditory processing. The activity of the superior colliculi may be a precursor of the mind and self processes that later blossom in the cerebral cortices. As for the nucleus tractus solitarius and parabrachial nucleus, they are the very first providers of whole-body maps to the central nervous system. The activity in those maps, as we shall see, corresponds to primordial feelings.

Mapping applies not only to visual patterns but to *every* kind of sensory pattern the brain is involved in constructing. For example, the mapping of sound begins in the ear's equivalent of the retina: the cochlea, located in our inner ear, one on each side. The cochlea receives the mechanical stimuli that result from the vibration of the tympanic membrane and of a small collection of bones located underneath it. The equivalent of the retinal neurons for the cochlea are the hair cells. At the top of a hair cell, a wisp of hair (the bundle) moves under the influence of sound energy and provokes an electrical current captured by the axon terminal of a neuron located in the cochlear ganglion. This neuron sends messages into the brain across six separate stations that form a chain—the cochlear nucleus, the superior olivary nucleus, the nucleus of the lateral lemniscus, the inferior colliculus, the medial geniculate nucleus, and finally the primary auditory cortex. The latter is comparable, in terms of hierarchy, to the primary visual cortex. The auditory cortex is the beginning of yet another signaling chain within the cerebral cortex itself.

The very first auditory maps are formed in the cochlea, just as the very first visual maps are formed in the retina. How are the sound maps achieved? The cochlea is a spiral ramp with an overall conical shape. It resembles a snail shell, as the Latin root of the word *cochlea* suggests. If you have ever been at the Guggenheim Museum in New York, you can easily picture what goes on inside the cochlea. All you need to do is imagine that the circles tighten as you go up and that the overall shape of the building is a cone pointing up. The ramp that you walk on wraps around the vertical axis of the cone, just like the cochlea's. Within the spiral ramp the hair cells are located with an exquisite ordering determined by the sound frequencies to which they are capable of responding. The hair cells that respond to the highest frequencies are located at the base of the cochlea, which means that as you ascend the ramp, the other frequencies follow in descending order until the apex of the cochlea is reached, as that is where the hair cells respond to the lowest frequencies. It all starts with lyric sopranos and ends with deep basses. The upshot is a spatial map of possible tones ordered by frequency, a tonotopic map. Remarkably, a version of this sound map is repeated at every one of the five subsequent stations of the auditory system on the way to the auditory cortex, where the map is finally laid out in a sheath. We hear an orchestra playing or the voice of a singer when neurons along the auditory chain become active and when the final cortical layout distributes spatially all the rich substructures of the sounds coming to our ears.

The mapping scheme applies far and wide to patterns having to do with body structure, such as a limb and its movement or the breakage in the skin caused by a burn, or to the patterns that result from touching the car keys you hold in your hand, surveying their shape and the smooth texture of their surface.

The closeness between mapped patterns in the brain and the actual objects that prompt them has been demonstrated in a variety of studies. For example, it is possible to uncover, in a monkey's visual cortex, a strong correlation between the structure of a visual stimulus (e.g., a circle or a cross) and the pattern of activity it evokes. This was first shown by Roger Tootell in brain tissue obtained from monkeys. However, under no circumstances can we "observe" the monkey's visual experience—the images the monkey itself sees. Images—visual, auditory, or of whatever other variety one may wish—are available *directly* but *only* to the owner of the mind in which they occur. They are private and unobservable by a third party. All the third party can do is guess.

Neuroimaging studies of the human brain are also beginning to uncover such correlations. Using multivariate pattern analysis, several research groups, ours included, have been able to show that certain patterns of activity in human sensory cortices correspond distinctively to a certain class of object.[3]

Maps and Minds

A spectacular consequence of the brain's incessant and dynamic mapping is the mind. The mapped patterns constitute what we, conscious creatures, have come to know as sights, sounds, touches, smells, tastes, pains, pleasures, and the like—in brief, images. The images in our minds are the brain's momentary maps of everything and of anything, inside our body and around it, concrete as well as abstract, actual or previously recorded in memory. The words I am using to bring these ideas to you were first formed, however briefly and sketchily, as auditory, visual, or somatosensory images of phonemes and morphemes, before I implemented them on the page in their written version. Likewise, the written words, now printed before your eyes, are first processed by you as *verbal* images (visual images of written language) before their action on the brain promotes the evocation of yet other images, of a *nonverbal* kind. The nonverbal kinds of images are those that help you display mentally the concepts that correspond to words. The feelings that make up the background of each mental instant and that largely signify aspects of the body state are images as well. Perception, in whatever sensory modality, is the result of the brain's cartographic skill.

Images represent physical properties of entities and their spatial and temporal relationships, as well as their actions. Some images, which probably result from the brain's making maps of itself making maps, are actually quite abstract. They describe patterns of occurrence of objects in time and space, the spatial relationships and movement of objects in terms of velocity and trajectory, and so forth. Some images find their way into musical compositions or mathematical descriptions. The process of mind is a continuous flow of such images, some of which correspond to actual, ongoing business outside the brain, while some are being reconstituted from memory in the process of recall. Minds are a subtle, flowing combination of actual images and recalled images, in ever-changing proportions. The mind's images tend to be logically interrelated, certainly when they correspond to events in the external world or in the body, which are, in and of themselves, governed by laws of physics and biology that define what we regard as logical. Of course, when you are daydreaming, you may produce illogical continuities of images, the same if you are having vertigo—the room does not really spin, the table is not turning on you, although the images tell you otherwise—and the same if you have taken a hallucinogenic drug. Such special situations apart, more often than not the flow of images moves forward in time, speedily or slowly, orderly or jumpily, and on occasion the flow moves along not just in one sequence but in several. Sometimes the sequences are concurrent, running in parallel; sometimes they intersect and become superposed. When the conscious mind is at its sharpest, the sequence of images is streamlined, barely letting us glimpse the surrounding fringes.

But in addition to the logic imposed by the unfolding of events in the reality external to the brain—a logical arrangement that the naturally selected circuitry of our brains foreshadows from the very early stages of development—the images in our minds are given more or less saliency in the mental stream according to their value for the individual. And where does that value come from? It comes from the original set of dispositions that orients our life regulation, as well as from the valuations that all images we have gradually acquired in our experience have been accorded, based on the original set of value dispositions during

our past history. In other words, minds are not just about images entering their procession naturally. They are about the cinemalike editing
choices that our pervasive system of biological value has promoted. The
mind procession is not about first come, first served. It is about value-
stamped selections inserted in a logical frame over time.[4]

Finally, and this is another critical issue, minds can be *either* nonconscious or conscious. Images continue to be formed, perceptually and in
recall, even when we are not conscious of them. Many images never get
the favors of consciousness and are not heard from, or seen directly, in
the conscious mind. And yet, in many instances, such images are capable
of influencing our thinking and our actions. A rich mental process
related to reasoning and creative thinking can proceed while we are
conscious of something else. I will return to the issues of the nonconscious mind in Part IV.

In conclusion, images are based on changes that occur in the body and
brain during the physical interaction of an object with the body. Signals
sent by sensors located throughout the body construct neural patterns
that map the organism's *interaction* with the object. The neural patterns
are formed transiently in the varied sensory and motor regions of the
brain that normally receive signals coming from specific body regions.
The assembling of the transient neural patterns is made from a selection
of neuron circuits recruited by the interaction. One can conceive of
those neuron circuits as preexisting building blocks within the brain.

Brain mapping is a distinctive functional feature of a system devoted
to managing and controlling the life process. The brain's mapping ability serves its managing purpose. At a simple level, the mapping may
detect the presence or provide the position of an object in space or the
direction of its trajectory. That may be helpful to track either a danger
or an opportunity, and either to avoid it or to seize it. And when our
minds avail themselves of multiple maps of every sensory variety and
create a multiplex perspective on the universe external to the brain, we
can respond to the objects and events in that universe with greater precision. Moreover, once maps are committed to memory and can be

brought back in imaginative recall, we are able to plan ahead and invent better responses.

The Neurology of Mind

Is it reasonable to ask which parts of the brain are mind-competent and which parts are not? The question is tricky but legitimate. A century and a half of research on the consequences of brain lesions provides the evidence we need to sketch a preliminary answer. Certain brain regions, in spite of their important contributions to major brain functions, are not involved in basic mind-making. Certain regions are definitely involved in making mind at a basic, indispensable level. And some other regions assist in mind-making, with tasks that involve the generation and regeneration of images, as well as management of image flow, such as editing of images and creation of continuities.

The entire spinal cord is apparently not essential to basic mind-making. The complete loss of the spinal cord results in severe motor defects, profound losses of body sensation, and some dampening of emotion and feeling. However, as long as the vagus nerve, which runs parallel to the spinal cord, is preserved (as it almost always is in such cases), the cross-signaling between brain and body remains robust enough to ensure autonomic control, to operate basic emotions and feelings, and to maintain the aspects of consciousness that require body input. Mind-making is definitely not obliterated by spinal cord damage, as we know so well from all the sad cases of people injured in accidents, at whatever level of the spinal cord the accident does its damage. Christopher Reeve's fine mind survived his extensive spinal cord damage, as did his consciousness. Outwardly, as I recall from meeting him, only the subtle operation of his emotional expressions had been slightly compromised. I suspect that mental representations of somatosensory stimuli from limbs and trunk are fully assembled only at the level of upper-brain-stem nuclei, with signals hailing from both spinal cord and vagus nerve, thus leaving the spinal cord in a peripheral position relative to basic mind-making. (Another way of positioning the spinal cord relative to mind-making is to say that its contributions are not missed by one's global function even if, when the contributions are present, they can be well appreciated. After spinal cord transections, patients will not feel pain but will show "pain-related" reflexes, indicating that the mapping of tissue injury is still being carried out at cord level but not signaled upward to the brain stem and cerebral cortex.)

The same dispensation applies to the cerebellum, certainly in the case of adults. The cerebellum plays important roles in the coordination of movement and the modulation of emotion, and it is involved in the learning and recall of skills and in cognitive aspects of skill development. But mind-making of the basic kind, as far as one can tell, is not its thing. We can say the same about the hippocampus, which is critical for learning new facts and which is regularly engaged by the normal process of recall but whose absence does not compromise basic mind-making. Both the cerebellum and the hippocampus are assistants to the editing and continuity processes, for images as well as movements, along with several cortical regions dedicated to motor control that probably play a role in assembling continuities in the mind process as well. This is critical, of course, for the comprehensive functioning of a mind, but it is not required for the basic making of images. The negative evidence regarding the mind-making capacities of the hippocampus and the adjacent cortices is especially powerful. It comes from the behaviors and self-reports of patients whose hippocampi and anterior temporal cortices are destroyed bilaterally, as a result of conditions such as anoxic injury, herpes simplex encephalitis, or surgical ablation. Their learning of new facts is largely precluded, as is, to a smaller or greater extent, the possibility of recalling the past. Yet the patients' minds are still immensely rich, with mostly normal perception in the visual, auditory, and tactile domains, and their recall of knowledge at generic (nonunique) levels is abundant. The fundamental aspects of their consciousness are largely intact.

When we turn to the cerebral cortex, the panorama is radically different. Several regions of the cerebral cortex are unequivocally involved in making the very images we behold and manipulate in our minds. And those cortices that do not make images tend to be involved in recording them or manipulating them in the process of reasoning, decision, and action. The early sensory cortices for vision, hearing, somatic sensation, taste, and smell, which appear like islands in the ocean of the cerebral cortex, certainly make images. These islands are aided in the task by thalamic nuclei of two kinds: relay nuclei (which bring inputs from the periphery) and associative nuclei (with which large sectors of the cerebral cortex are bidirectionally connected).

Powerful evidence supports this claim. We know that significant damage to each island of sensory cortex extensively disables the mapping function of that particular sector. For example, victims of bilateral damage to the early visual cortices become "cortically blind." Patients so affected are no longer able to form detailed visual images, not just in perception but often in recall as well. They may be left with a residual so-called blindsight, in which nonconscious clues permit some visual guidance of actions. A comparable situation applies to situations of significant damage to other sensory cortices. The remainder of the cerebral cortex, the ocean around the islands, albeit not primarily involved in making images, participates in the construction and processing of images, that is, in the recording, recalling, and manipulating of images generated in early sensory cortices, which are discussed in Chapter 6.[5]

But contrary to tradition and convention, I believe that the mind is not made in the cerebral cortex alone. Its first manifestations arise in the brain stem. The idea that mind processing begins at brain-stem level is so unconventional that it is not even unpopular. Among those who have championed the idea with great passion, I single out Jaak Panksepp. This idea, and that of early feelings arising in the brain stem, are of a piece.[6] Two brain-stem nuclei, the nucleus tractus solitarius and the parabrachial nucleus, are involved in generating basic aspects of the mind, namely, the feelings generated by ongoing life events, which

Varieties of Maps (images)	Source Objects
I maps of the organism's internal structure and state (interoceptive maps)	the functional condition of body tissues such as the degree of contraction / distension of smooth musculature; parameters of internal milieu state
II maps of other aspects of the organism (proprioceptive maps)	images of specific body components such as joints, striated musculature, some viscera
III maps of the world external to the organism (exteroceptive maps)	any object or event that engages a sensory probe such as the retina, the cochlea, or the mechano-receptors of the skin

Figure 3.1: Varieties of maps (images) and their source objects. When maps are experienced, they become images. A normal mind includes images of all three varieties described above. Images of an organism's internal state constitute *primordial feelings*. Images of other aspects of the organism combined with those of the internal state constitute specific *body feelings*. Feelings of emotions are variations on complex body feelings caused by and referred to a specific object. Images of the external world are normally *accompanied by* images of varieties of I and II.

Feelings are a variety of image, made special by their unique relation to the body (see Chapter 4). Feelings are spontaneously *felt* images. All other images are felt because they are accompanied by the particular images we call feelings.

include those described as pain and pleasure. I envision the maps generated by these structures as simple and largely devoid of spatial detail, but they result in feelings. These feelings are, in all likelihood, the primordial constituents of mind, based on direct signaling from the body proper. Interestingly, they are also primordial and indispensable components of the self and constitute the very first and inchoate revelation, to the mind, that *its* organism is alive.

These important brain-stem nuclei do not produce mere virtual maps of the body; they produce *felt* body states. And if pain and pleasure *feel* like something, these are the structures we first have to thank, along with the motor structures with which they incessantly loop back to the body, namely, those of the periaqueductal gray nuclei.

The Beginnings of Mind

To illustrate what I mean when I talk about the beginnings of mind, I need to discuss, however briefly, three sources of evidence. One comes from patients whose insular cortices have been damaged. Another comes from children born without a cerebral cortex. The third has to do with the functions of the brain stem in general and the functions of the superior colliculi in particular.

FEELING PAIN AND PLEASURE AFTER INSULAR DESTRUCTION

In the chapter on emotions (Chapter 5) we shall see that the insular cortices are unequivocally involved in the processing of a large range of feelings, from those that follow emotions to those that signify pleasure or pain, known as bodily feelings for short. Unfortunately, the powerful evidence relating feelings to the insula has been taken to mean that the substrate of all feelings is to be found only at the cortical level; the insular cortices thus pose as the rough equivalent of the early visual and auditory cortices. But just as the destruction of visual and auditory cortices does not abolish vision and hearing, the complete destruction of the insular cortices, from front to back, in both left and right cerebral hemispheres, does not result in a complete abolition of feeling. On the contrary, feelings of pain and pleasure remain after damage to *both* insular cortices caused by herpes simplex encephalitis. Along with my colleagues Hanna Damasio and Daniel Tranel, I have repeatedly observed that these patients respond with pleasure or pain to a variety of stimuli and continue feeling emotions, which they unequivocally report. Patients report discomfort with temperature extremes; they are displeased by boring tasks and are annoyed when their requests are refused. The social reactivity that depends on the presence of emotional feelings is not compromised. Attachment is maintained even to persons who cannot be recognized as loved ones and friends because, as part of the herpetic syndrome, concomitant damage to the anterior sector of the temporal lobes severely compromises autobiographical memory. Moreover, experimental manipulation of stimuli leads to demonstrable changes in the experience of feelings.[7]

It is reasonable to propose that in the absence of both insular cortices, the feelings of pain and pleasure arise in two brain-stem nuclei I mentioned earlier (the tractus solitarius and the parabrachial), both of which are suitable recipients of signals from the body's interior. In normal individuals, these two nuclei send their signals on to the insular cortex via dedicated nuclei of the thalamus (Chapter 4). In brief, whereas the brain-stem nuclei would ensure a basic level of feelings, the insular cortices would provide a more differentiated version of those feelings and, most important, would be able to relate the feelings to other aspects of cognition based on activity elsewhere in the brain.[8]

The circumstantial evidence in favor of this idea is telling. The nucleus tractus solitarius and the parabrachial nucleus receive a full complement of signals describing the state of the internal milieu in the entire body. Nothing escapes them. There are signals from the spinal cord and trigeminal nucleus, and even signals from "naked" brain regions such as the nearby area postrema, that are devoid of the protective blood-brain barrier and whose neurons respond directly to molecules traveling in the bloodstream. The signals compose a comprehensive picture of the internal milieu and viscera, and that picture happens to be the prime component of our feeling states. These nuclei are

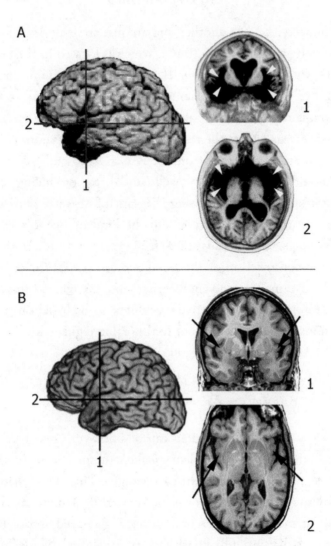

Figure 3.2: Panel A shows the MR scan of a patient with complete damage to the insular cortices, in both the left and right hemispheres. A three-dimensional reconstruction of the patient's brain is shown on the left. On the right, there are two sections taken through the brain (marked 1 and 2), along the vertical and horizontal black lines shown on the left and marked respectively 1 and 2. The area shown in black corresponds to brain tissue destroyed by the disease. The white arrows point to locations where the insula should have been. Panel B shows a normal brain in 3-D and in two sections taken at the same levels. The black arrows point to the normal insular cortex.

richly connected to one another and are just as richly interconnected
with the periaqueductal gray (PAG), which is located in their vicinity.
The PAG is a complex set of nuclei, with multiple subunits, and is the
originator of a large range of emotional responses related to defense,
aggression, and coping with pain. Laughter and crying, expressions of
disgust or fear, as well as the responses of freezing or running in situations of fear are all triggered from the PAG. The to-and-fro of connections among these nuclei is well suited to producing complex
representations. The basic wiring diagram of these regions qualifies
them for an image-making role, and the kind of image these nuclei
make is feelings. Also, because these feelings are early and foundational
steps in the construction of the mind and are critical for the maintenance of life, it makes good engineering sense (by which I mean evolutionary sense) for the supportive machinery to be based on structures
that are housed literally next door to those that regulate life.[9]

THE STRANGE SITUATION OF CHILDREN
DEPRIVED OF THE CEREBRAL CORTEX

For a variety of reasons, children can be born with intact brain-stem
structures but largely absent telencephalic structures, namely, the cerebral cortex, the thalamus, and the basal ganglia. This unfortunate condition is commonly due to a major stroke, occurring in utero, as a result of
which all or most of the cerebral cortex is damaged and reabsorbed,
leaving the skull cavity filled with cerebrospinal fluid. This is known as
hydranencephaly, to distinguish it from developmental defects, generally known as anencephaly, that compromise other structures beside the
cerebral cortex.[10] The affected children can survive for many years, even
past adolescence, and are often dismissed as "vegetative." They are commonly institutionalized.

These children, however, are anything but vegetative. On the contrary, they are awake and behaving. To a limited but by no means negligible extent, they can communicate with their caregivers and interact

with the world. They are patently *minded* in a way that patients in vegetative state or akinetic mutism are not. Their misfortune provides a rare window into the sort of mind that can still be engendered when the cerebral cortex is absent.

What do these unfortunate children look like? Their motions are quite limited by the lack of muscular tone in their spine and the spasticity of their limbs. But they move their heads and eyes freely, they have expressions of emotion in their faces, they can smile at stimuli that one would expect a normal child to smile at—a toy, a certain sound—and they can even laugh and express normal joy when they are tickled. They can frown and withdraw from painful stimuli. They can move toward an object or situation they crave—for example, crawl toward a spot on the floor where sunlight is falling and where the child will bask in the sun and obviously draw benefit from the warmth. The children *look* pleased, in an external manifestation of the kind of feelings one would predict they would have following an emotional response appropriate to the stimulus.

These children can orient head and eyes, albeit inconsistently, to the person addressing them or touching them and reveal preferences for distinct people. They tend to be fearful of strangers and appear happiest near their habitual mother/caregiver. Likes and dislikes are apparent, none so striking as in examples of music. The children tend to like some musical pieces more than others; they can respond to different instrumental sounds and different human voices. They also can respond to different tempi and different composition styles. Their faces are a good reflection of their states of emotion. In brief, they are most joyful when they are touched and tickled, when preferred music pieces are played, and when certain toys are shown in front of their eyes. Obviously they hear and they see, although we have no way of knowing how well. Their hearing seems superior to their sight.

Of necessity, whatever they see and hear is achieved subcortically, in all likelihood in the colliculi, which are intact. Whatever they feel is achieved subcortically by the nucleus tractus solitarius and parabrachial nucleus, which are intact, as they have no insular cortex or somatosensory cortices I and II to assist with such a task. The emotions they produce must be triggered from the nuclei in the periaqueductal gray and must be executed by the cranial nerve nuclei that control facial expressions of emotions (those nuclei are also intact). The running of the life process is supported by an intact hypothalamus, located immediately above the brain stem and helped by an intact endocrine system and by the vagus nerve network. Hydranencephalic girls even develop menstrual periods at puberty.

That these children give some evidence of mind process is not in doubt. Likewise, their expressions of joy, sustained as they are over many seconds and even minutes, and consonant as they are with the causative stimulus, can be reasonably associated with feeling states. It is compelling for me to assume that the *delight* they exhibit is real *felt* delight, even if they cannot report it in so many words. That being so, they would achieve the bottom riser of a stepwise mechanism leading to consciousness, namely, feelings connected to an integrated representation of the organism (a protoself), possibly modified by object engagement, constituting an elementary experience.

The possibility that they do have a conscious mind, albeit an extremely modest one, is supported by an intriguing finding. When these children suffer an absence seizure, the caregivers easily detect its onset; they can also tell when the seizure ends and report that the "child is returned to them." The seizure appears to suspend the minimal consciousness they normally exhibit.

Hydranencephalic individuals present a most troubling picture, one that informs us of the limits, in humans, of both brain-stem structures and cerebral cortex. The condition gives the lie to the claim that sentience, feelings, and emotions arise only out of the cerebral cortex. That cannot possibly be the case. The degree of sentience, feeling, and emotion that is possible in these cases is quite limited, of course, and, most important, disconnected from the wider world of mind that, indeed, only the cerebral cortex can provide. But having spent a good part of my life studying the effects of brain damage on the human mind and behavior, I can say that these children have little in common with patients in vegetative state, a condition in which the interaction with the world is even more reduced and that can actually be caused by damage to precisely the same regions of the brain stem that *are* intact in hydranencephalics. If a parallel could be drawn at all, once the motor defects are factored out, it would be between hydranencephalic children and newborn infants, in which a mind is clearly at work but where the core self is barely beginning to gather. This is in keeping with the fact that hydranencephalics may be first diagnosed months after birth, when parents note a failure to thrive and scans reveal a catastrophic absence of cortex. The reason behind the vague similarity is not too difficult to fathom: normal infants lack a fully myelinated cerebral cortex, which still awaits development. They already have a functional brain stem but only a partially functional cerebral cortex.

A NOTE ON THE SUPERIOR COLLICULUS

The superior colliculi are part of the tectum, a region that is closely interrelated with the periaqueductal gray nuclei and, indirectly, with the nucleus tractus solitarius and parabrachial nucleus. The involvement of the superior colliculus in visual-related behavior is well known. But the possible role of these structures in the process of mind and self is rarely considered, although there are notable exceptions in the work of Bernard Strehler, Jaak Panksepp, and Bjorn Merker.[11] The anatomy of the superior colliculus is fascinating and all but compels us to guess what this structure is supposed to accomplish. The superior colliculus has seven layers; layers I through III are the "superficial" layers, while layers IV through VII are called "deep." All the connections coming to and going out of the superficial layers have to do with vision, and layer II, the main superficial layer, receives signals from the retina and from the primary visual cortex. These superficial layers assemble a retinotopic map of the contralateral visual field.[12]

The deep layers of the superior colliculus contain, in addition to a map of the visual world, topographical maps of auditory and somatic information, the latter hailing from the spinal cord as well as the hypothalamus. The three varieties of maps—visual, auditory, and somatic—are in spatial register. This means that they are stacked in such a precise way that the information available in one map for, say, vision, corresponds to the information on another map that is related to hearing or body state.[13] There is no other place in the brain where information available from vision, hearing, and multiple aspects of body states is so literally superposed, offering a prospect of efficient integration. The integration is made all the more significant by the fact that its results can gain access to the motor system (via the nearby structures in the periaqueductal gray as well as via the cerebral cortex).

The other day a nice little lizard on my terrace was darting about in hot pursuit of a silly fly that insisted on buzzing him, flying dangerously low. The lizard tracked the fly perfectly and finally caught it with its tongue, thrown out at the precise moment. The collicular neurons plotted the moment-to-moment position of the fly and guided the lizard's muscles accordingly, eventually dispatching the tongue when the prey was within reach. The adaptive perfection of this visuomotor behavior to its environment is astounding. But now imagine the rapid, sequential firing of the neurons in the lizard's superior colliculus, astound yourself some more, and pause for a second to wonder. What did the lizard *see*? I would not know for certain, but I suspect he saw a black moving dot, zigzagging in an otherwise vague field of vision. What did the lizard *know* of the ongoing event? I suspect nothing, in our sense of knowing. And what did he feel when he was eating his hard-won lunch? I suspect that his brain stem registered the successful completion of its goal-directed behavior and the results of an improved state of homeostasis. The substrates of the lizard's feelings were probably in place, although he could not reflect on the remarkable skill he had just displayed. It is not always easy being green.

This powerful integration of signals serves an obvious and immediate purpose: the gathering of information necessary to guide effective action, be it movement of the eyes, the limbs, or even the tongue. This is achieved by rich connections from the colliculi to all the brain regions required to guide movement effectively, in the brain stem, in the spinal cord, in the thalamus, and in the cerebral cortex. But besides achieving effective guidance of movement, it is possible that there are "internal," mental consequences of this useful arrangement. In all likelihood, the integrated, in-register maps of the superior colliculus generate images as well—nowhere as rich as those made in the cerebral cortex, but images nonetheless. Some of the beginnings of mind are probably to be found here, and the beginnings of self might be found here too.[14]

What about the superior colliculus in humans? In humans, selective destruction of the superior colliculus is rare, so rare that the neurologic literature records a single case, of bilateral damage, fortunately studied by a major neurologist and neuroscientist, Derek Denny-Brown.[15] The lesion was the result of trauma, and the patient survived for months, in a severely impaired state of consciousness that best resembled an akinetic mute state. This is suggestive of a compromise of mentation, but I must add that on the occasion I encountered a patient with collicular damage, only a brief disturbance of consciousness was detectable.

Seeing with the colliculi alone once the visual cortices are lost possibly consists of sensing that some unspecified object X is moving in one of the quadrants of vision, say, away from me, or that it is coming closer to me. In neither case will I be able to describe what the object is mentally, and I may not even be conscious of it. We are talking here of a very vague mind, gathering sketchy information about the world, although the fact that the images are vague and incomplete does not render them useless or unhelpful, as blindsight shows. However, when the visual cortices are missing from birth, as in the hydranencephalic cases described earlier, both the superior and the inferior colliculi may make more substantial contributions to the mind process.

I must add one last fact to the evidence in favor of promoting the superior colliculi to mind-contributing status. The superior colliculus produces electrical oscillations in the gamma range, a phenomenon that has been linked to synchronic activation of neurons and that has been proposed by the neurophysiologist Wolf Singer to be a correlate of coherent perception, possibly even of consciousness. To date, the superior colliculus is the only brain region outside the cerebral cortex known to exhibit gamma-range oscillations.[16]

Closer to the Making of Mind?

The picture that emerges from the foregoing indicates that mind-making is a highly selective business. It is not the case that the entire central nervous system is uniformly involved in the process. Certain regions are not involved, some are involved but are not principal players, and some carry out the bulk of the work. Among the last, some provide detailed images; others provide a simple but foundational kind of images such as bodily feelings. All regions involved in mind-making have highly differentiated patterns of interconnectivity, suggestive of very complex signal integration.

Contrasting the set of regions that do and do not contribute to the mind-making effort does not tell us what kind of signals neurons must produce; it does not specify frequencies or intensities of neuron firing or patterns of coalition among neuron sets. But it tells about certain aspects of the wiring diagram that neurons require to be involved in mind-making. For example, the cortical mind-making sites are clusters of interlocked regions organized around the port of entry for inputs from peripheral sensory probes. The subcortical mind-making sites are also intensely interlocked clusters of regions, nuclei in this case, and they are also organized around inputs from another "periphery"— namely, the body itself.

Another requirement, applying equally to cerebral cortex and subcortical nuclei: there must be massive interconnectivity among the mind-making regions so that recursiveness is prevalent and a high complexity of cross-signaling is achieved, a feature that in the case of the cortex is amplified by corticothalamic interlocking. (The terms *reentrant* and *recursive* refer to signaling that, rather than merely going forward along a single chain, also returns to the origin, looping back to the ensemble of neurons where each element of the chain begins.) Mind-making regions in the cortex also receive numerous inputs from a variety of nuclei located underneath, some in the brain stem and some in the thalamus; they modulate cortical activity by way of neuromodulators (such as catecholamines) and neurotransmitters (such as glutamate).

Finally, a certain timing of the signaling is necessary so that elements of a stimulus that arrive together at the peripheral sensory probe can stay together as the signals are being processed within the brain. For mind states to emerge, small circuits of neurons must behave in a very particular manner. For example, in small circuits whose activity signifies that a certain feature is present, neurons increase their firing rates. Ensembles of neurons that are working together to signify some combination of features must *synchronize* their firing rates. This was first demonstrated in the monkey by Wolf Singer and his colleagues (and also by R. Eckhorn), who found that separate regions of the visual cortex involved in processing the same object exhibited synchronized activity in the 40 Hz range.[17] The synchronization is probably achieved by oscillations of neuronal activity. When brains are forming perceptual images, the neurons of the separate regions that contribute to the percept exhibit synchronized oscillations in the high-frequency gamma range. This could be part of the secret behind the "binding" of separate regions by means of time; I will invoke this sort of mechanism to explain the operation of convergence-divergence zones (Chapter 6) and the assembly of the self (Chapters 8, 9, and 10).[18] In other words, besides building rich maps at a variety of separate locations, the brain must *relate* the maps to one another, in coherent ensembles. Timing may well be the key to relating.

In sum, the notion of a map as a discrete entity is merely a helpful abstraction. The abstraction hides the extremely large number of neuron interconnections that are involved in each separate region and that generate a huge degree of signaling complexity. What we experience as mental states corresponds not just to activity in a discrete brain area but rather to the result of massive recursive signaling involving multiple regions. And yet, as I will argue in Chapter 6, the explicit aspects of certain mind contents—a specific face, a certain voice—are likely to be *assembled* within a particular collection of brain regions whose design lends itself to map assembly, albeit with the help of other contributing regions. In other words, there is some anatomical specificity behind the making of mind, some fine functional differentiation within the maelstrom of global neuron complexity.

As one struggles to understand the neural basis of mind, one may well ask if the foregoing is good news or bad. There are two ways of responding to that question. One way is to feel somewhat discouraged by so much booming, buzzing confusion and despair that a clear, well-lighted pattern can ever be gleaned from the biological mess. But one might also embrace complexity wholeheartedly and realize that the brain needs the seeming mess in order to generate something as rich, smooth, and adaptive as mental states. I choose the second option. I would have a hard time believing that a discrete map in a single cortical region could, by itself, ever allow me to hear the Bach piano partitas or behold Venice's Grand Canal, let alone enjoy them and discover their significance in the large scheme of things. As far as the brain is concerned, less is more only when we wish to communicate the gist of a phenomenon. Otherwise, more is always better.

4

The Body in Mind

The Topic of the Mind

Before consciousness came to be regarded as the central problem in mind and brain research, a closely related issue, known as the mind-body problem, dominated the intellectual debate. In one form or another, it permeated the thinking of philosophers and scientists from Descartes and Spinoza to the present. The functional arrangement described in Chapter 3 makes my position on this problem clear: the brain's map-making ability provides an essential element in its solution. In brief, complex brains such as ours naturally make explicit maps of the structures that compose the body proper, in more or less detail. Inevitably brains also map the functional states naturally assumed by those body components. Because, as we have seen, brain maps are the substrate of mental images, map-making brains have the power of literally introducing the body as *content* into the mind process. Thanks to the brain, the body becomes a natural topic of the mind.

But this body-to-brain mapping has a peculiar and systematically overlooked aspect: although the body is the thing mapped, it never loses contact with the mapping entity, the brain. Under normal circumstances they are hitched to each other from birth to death. Just as important, the mapped images of the body have a way of permanently influencing the very body they originate in. This situation is unique. It has no parallel in the mapped images of objects and events external to the body, which can never exert any direct influence on those objects and events. I believe that any theory of consciousness that does not incorporate these facts is doomed to fail.

The reasons behind the body-to-brain connection have been presented already. The business of managing life consists of managing a body, and the management gains precision and efficiency from the presence of a brain—specifically, from having circuits of neurons assisting the management. I said that neurons are about life and about managing life in other cells of the body, and that that *aboutness* requires two-way signaling. Neurons act on other body cells, via chemical messages or excitation of muscles, but in order to do their job, they need inspiration from the very body they are supposed to prompt, so to speak. In simple brains, the body does its prompts simply by signaling to subcortical nuclei. Nuclei are filled with "dispositional know-how," the sort of knowledge that does not require detailed mapped representations. But in complex brains, the map-making cerebral cortices describe the body and its doings in so much explicit detail that the owners of those brains become capable, for example, of "imaging" the shape of their limbs and their position in space, or the fact that their elbows hurt or their stomach does. Bringing the body to mind is the ultimate expression of the brain's intrinsic aboutness, its *intentional* attitude regarding the body, to phrase it in terms that connect with the ideas of philosophers such as Franz Brentano.[1] Brentano actually saw the intentional attitude as the hallmark of mental phenomena and believed that physical phenomena lacked intentional attitudes and aboutness. This does not seem to be the case. As we saw in Chapter 2, single cells also *appear* to have intentions and aboutness in much the same sense. In other words, neither a whole brain nor single cells deliberately *intend* anything with their behavior, but their stance is as if they do. This is one more reason to deny the intuitive abyss between the mental and the physical worlds.[2] On this count, at least, there certainly is not one.

The brain's aboutness vis-à-vis the body has two other spectacular consequences, and they too are indispensable to resolving both the mind-body and consciousness conundra. The pervasive, exhaustive mapping of the body covers not only what we usually regard as the body proper—the musculoskeletal system, the internal organs, the internal milieu—but also the special devices of perception placed at specific sites of that body, the body's spying outposts—the smell and taste mucosae, the tactile elements of the skin, the ears, the eyes. Those devices are located in the body as much as the heart and guts are, but they occupy privileged positions. Let's say they are set like diamonds in a frame. All those devices have a part made of "old flesh" (the armature for the diamonds) and another made of the delicate and special "neural probe" (the diamonds). Important examples of old-flesh armature include the external ear, the ear canal, the middle ear with its ossicles and the tympanic membrane; and the skin and muscles around the eyes, and the varied components of the eyeball besides the retina, such as the lens and the pupil. Examples of the delicate neural probes include the cochlea in the inner ear, with its sophisticated hair cells and sound-mapping capabilities; and the retina at the back of the eyeball, onto which optical images are projected. The combination of old flesh and neural probe constitutes a body border. Signals hailing from the world must cross that border in order to enter the brain. They cannot simply enter the brain directly.

Because of this curious arrangement *the representation of the world external to the body can come into the brain only via the body itself,* namely via its surface. The body and the surrounding environment interact with each other, and the changes caused *in the body* by that interaction are mapped in the brain. It is certainly true that the mind learns of the outside world via the brain, but it is equally true that the brain can be informed only via the body.

The second special consequence of the brain's body aboutness is no less notable: by mapping its body in an integrated manner, the brain manages to create the critical component of what will become the self. We shall see that body mapping is a key to the elucidation of the problem of consciousness.

Finally, as if the above facts were not quite extraordinary, the close relationships of body and brain are essential to understanding something else that is central to our lives: spontaneous bodily feelings, emotions, and emotional feelings.

Body Mapping

How does the brain accomplish the mapping of the body? By treating the body proper and its parts as any other object, one might say, but that would hardly do justice to the problem, because as far as the brain is concerned, the body proper is more than just any object: it is the *central* object of brain mapping, the very first focus of its attentions. (Whenever I can, I use the term *body* to mean "body proper" and leave aside the brain. The brain is also part of the body, of course, but it has a particular status: it is the body part that can communicate to every other body part and toward which every other body part communicates.)

William James had an inkling of the extent to which the body needed to be brought to mind, but he could not know how intricate the mechanisms responsible for bringing about the body-to-mind transfer would turn out to be.[3] The body uses both chemical signals and neural signals to communicate with the brain, and the range of conveyed information is broader and more detailed than he could have envisioned. In effect, I am now convinced that talking merely about body-to-brain communication misses the point. Although part of the signaling from body to brain results in a straightforward mapping (for example, the mapping of the position of a limb in space), a substantial part of the signaling is first *treated* by subcortical nuclei, within the spinal cord and especially in the brain stem, which should not be conceived of as way stations for body signals en route to the cerebral cortex. As we shall see in the next section, something is added at that intermediate stage. This is quite important when it comes to the signals related to the body's interior that come to constitute feelings. Moreover, aspects of the body's physical structure and function are engraved in brain circuitry, from early in development, and generate persistent patterns of activity. In other words, some version of the body is permanently re-created in brain activity. The heterogeneity of the body is mimicked in the brain, one of the high marks of the brain's body-aboutness. Last, the brain can do more than merely map states that are actually occurring, with more or less fidelity: it can also *transform* body states and, most dramatically, *simulate* body states that have not yet occurred.

Those who are unacquainted with neuroscience may assume that the body operates as a single unit, a single lump of flesh connected to the brain by live wires called nerves. The reality is quite different. The body has numerous separate compartments. To be sure, the viscera to which so much attention is paid are essential. The incomplete list of viscera includes the usual suspects: the heart, the lungs, the gut, the liver and pancreas, the mouth, the tongue, and the throat; the endocrine glands (e.g., pituitary, thyroid, adrenals); the ovaries and testes. But the list needs also to include less-usual suspects: an equally vital but less-recognized organ, the skin, which envelops our entire organism; the bone marrow; and two dynamic shows called blood and lymph. All of these compartments are indispensable for the body's normal operation.

Perhaps not surprisingly, early human minds, less integrated and sophisticated than ours, easily perceived the broken-down, piecemeal reality of our bodies, as suggested by the words that have come to us from Homer. *Iliad* humans do not speak of a whole body (*soma*) but rather of body parts, namely, limbs. Blood and breath and visceral functions are designated by the word *psyche,* not yet called to duty as "mind" or "soul." The animation that drives the body, probably mixed with drive and emotion, is the *thumos* and the *phren*.[4]

Body-brain communication goes both ways, from body to brain and in reverse. The two ways of communication, however, are hardly symmetrical. The body-to-brain signals, neural and chemical, permit the brain to create and maintain a multimedia documentary on the body, and allow the body to alert the brain to important changes occurring in its structure and state. The internal milieu—the bath that all body cells inhabit and of which the blood chemistries are an expression—also sends signals to the brain, not via nerves but via chemical molecules, which impinge directly on certain parts of the brain designed to receive their messages. So the range of information conveyed to the brain is extremely wide. It includes, for example, the state of contraction or dilation of smooth muscles (the muscles that form, for example, the walls of the arteries, the gut, and the bronchi); the amount of oxygen and carbon dioxide concentrated locally in any region of the body; the temperature and the pH at various locations; the local presence of toxic chemical molecules; and so forth. In other words, the brain knows what the past state of the body has been and can be told of modifications occurring in that state. The latter is essential if the brain is to produce corrective responses to changes that threaten life. The brain-to-body signals, on the other hand, neural as well as chemical, consist of commands to change the body. The body tells the brain: this is how I am built and this is how you should see me now. The brain tells the body what to do to maintain its even keel. Whenever it is called for, it also tells the body how to construct an emotional state.

There is more to the body, however, than internal organs and internal milieu. There are also muscles, and they come in two varieties: smooth and striated. The striated variety shows characteristic "bands" (striae) under the microscope, while the smooth variety does not. Smooth muscles are evolutionarily ancient and are confined to viscera—our gut and our bronchi contract and distend thanks to smooth muscles. The walls of our arteries are made in good part of smooth muscles—one's blood pressure rises when they tighten their grip around the artery. Striated muscles, by contrast, are attached to bones in the skeleton and produce external body movement. The only exception to this scheme of things is the heart, which is also made of striated muscle fibers and whose contractions serve not body movement but rather the pumping of blood. Signals describing the state of the heart are sent to brain sites dedicated to the viscera, not to those that pertain to movement.

When skeletal muscles are connected to two bones articulated by a joint, the shortening of their fibers generates movement. Picking up an object, walking, talking, breathing, and eating are all actions that depend on the contraction and distension of skeletal muscles. Whenever those contractions occur, the configuration of the body changes. Except for moments of complete immobility, which are infrequent in the awake state, the configuration of the body in space changes continuously, and the map of the body represented in the brain changes accordingly.

In order to control movement with precision, the body must instantly convey to the brain information on the state of skeletal muscle contraction. This requires efficient nerve pathways, which are evolutionarily more modern than those that convey signals from the viscera and internal milieu. These pathways arrive in brain regions dedicated to sensing the state of these muscles.

As noted, the brain also sends messages to the body. In fact, many aspects of the body states being continuously mapped in the brain were caused in the first place by brain signals to the body. As in the case of communication from the body to the brain, the brain talks to the body via both neural and chemical channels. The neural channel uses nerves, whose messages lead to the contraction of muscles and the execution of actions. The chemical channels involve hormones, such as cortisol, testosterone, and estrogen. The release of hormones changes the internal milieu and the operation of the viscera.

Body and brain are engaged in a continuous interactive dance. Thoughts implemented in the brain can induce emotional states that are implemented in the body, while the body can change the brain's landscape and thus the substrate for thoughts. The brain states, which correspond to certain mental states, cause particular body states to occur; body states are then mapped in the brain and incorporated into the ongoing mental states. A small alteration on the brain side of the system can have major consequences for the body state (think of the release of any hormone); likewise, a small change on the body side (think of a broken tooth filling) can have a major effect on the mind once the change is mapped and perceived as acute pain.

From Body to Brain

The remarkable European school of physiology that flourished from the middle of the nineteenth century to the early twentieth described the contours of body-to-brain signaling with admirable accuracy, but the relevance of this general scheme for the understanding of the mind-body problem went unnoticed. The neuroanatomical and neurophysiological details, not surprisingly, have been uncovered only in the past few years.[5]

The state of the body's *interior* is conveyed to the brain via dedicated neural channels to specific brain regions. Special nerve-fiber types (Aδ and C fibers) bring signals from every nook and cranny of the body into selected parts of the central nervous system (such as the lamina-I section of the posterior horn of the spinal cord), at every level of the vertical spinal cord length, and the pars caudalis of the trigeminal nerve. The spinal cord components handle signals from the internal milieu and viscera of the body except the head—the chest, abdomen, and limbs. The trigeminal nerve nucleus handles signals from the internal milieu and viscera of the head, including the face and its skin, the scalp, and the paramount pain-generating meningeal membrane, the dura mater. Equally dedicated are the brain regions charged with handling the signals after they enter the central nervous system and as subsequent signals march toward the higher levels of the brain.

The least one can say is that along with chemical information available in the bloodstream, these neural messages inform the brain about the state of a good part of the body's interior—the state of the visceral-chemical body components beneath the skin's outer perimeter.

Complementing the complex mapping of the interior sense described above, to which we refer as *interoception,* are the body-to-brain channels that map the state of skeletal muscles engaged in movement, which are a part of *exteroception*. Messages from the skeletal muscles use different and fast-conducting kinds of nerve fibers—$A\alpha$ and $A\gamma$ fibers—as well as different stations of the central nervous system all the way into the higher levels of the brain. The upshot of all this signaling is a multidimensional picture of the body in the brain and, thus, in the mind.[6]

Representing Quantities and Constructing Qualities

The body-to-brain signaling I have described does not deal merely with the representation of quantities of certain molecules or degrees of smooth muscle contraction. To be sure, the body-to-brain channels do transmit information regarding quantities (how much CO_2 or O_2 is present; how much sugar is in the blood; and so forth). But there is, side by side, a *qualitative aspect* to the results of the transmission. The state of the body is felt to be in some variation of pleasure or pain, of relaxation or tension: there can be a sense of energy or lassitude, of physical lightness or heaviness; of unimpeded flow or resistance, of enthusiasm or discouragement. How can this qualitative background effect be achieved? To begin with, by arranging the varied quantitative signals

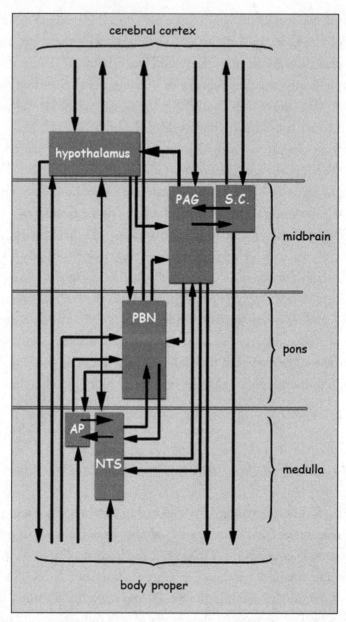

Figure 4.1: Schematics of key brain-stem nuclei involved in life regulation (homeostasis). Three brain-stem levels are marked in descending order (midbrain, pons, and medulla); the hypothalamus (which is a functional component of the brain stem even if it is, anatomically, a part of the diencephalon) is also included. Signaling to and from the

body proper and to and from the cerebral cortex is indicated by vertical arrows. Only the basic interconnections are depicted, and only the main nuclei involved in homeostasis are included. The classic reticular nuclei are not included, nor are the monoaminergic and cholinergic nuclei.

The brain stem is often considered a mere conduit for signals from body to brain and brain to body, but the reality is different. Structures such as the NTS (nucleus tractus solitarius) and PBN (parabrachial nucleus) do transmit signals, from body to brain but not passively. These nuclei, whose topographic organization is a precursor of that of the cerebral cortex, *respond* to body signals, thereby regulating metabolism and guarding the integrity of body tissues. Moreover, their rich, recursive interactions (signified by mutual arrows) suggest that in the process of regulating life, new patterns of signals can be created. The PAG (periaqueductal gray), a generator of complex chemical and motor responses aimed at the body (such as the responses involved in reacting to pain and in the emotions), is also recursively connected to the PBN and the NTS. The PAG is a pivotal link in the body-to-brain resonant loop.

It is reasonable to hypothesize that in the process of regulating life the networks formed by these nuclei also give rise to composite neural states. The word *feelings* describes the mental aspect of those states.

arriving in brain-stem structures and in insular cortices so as to *compose* diverse landscapes for the ongoing body events.

To grasp what I have in mind, I ask the reader to imagine a state of pleasure (or anguish) and try to itemize its components by making a brief inventory of the varied parts of the body that are changed in the process: endocrine, cardiac, circulatory, respiratory, intestinal, epidermic, muscular. Now consider that the feeling you will experience is the integrated perception of all such changes as they occur in the landscape of the body. As an exercise, you can actually try to *compose* the feeling and assign values of intensity to each component. For each instance that you imagine, you will obtain a different quality.

But there are other ways of constructing qualities. First, as noted earlier, a significant portion of body signals undergoes additional treatment within certain nuclei of the central nervous system. In other words, the signals are processed at intermediate stages, which are not mere relay stations. The machinery of emotion located in the nuclei of the periaqueductal gray is likely to influence processing of body signals at the level of the parabrachial nucleus, directly and indirectly. Exactly what is added in the process is not known, in neural terms, although the addition is likely to contribute to the experiential quality of feelings. Second, the regions that receive body-to-brain signaling respond, in turn, by altering the ongoing state of the body. I envision these responses as initiating a tight two-way, resonant loop between body states and brain states. The brain mapping of the body state and the actual body state are never far apart. Their border is blurred. They become virtually fused. The sense that events are occurring in the flesh would arise from this arrangement. A wound that is mapped in the brain stem (within the parabrachial nucleus), and that is perceived as pain, unleashes multiple responses back to the body. The responses are initiated by the parabrachial nucleus and executed in the nearby periaqueductal gray nuclei. They cause an emotional reaction and a change in the processing of subsequent pain signals, which immediately alter the body state and, in turn, alter the next map that the brain will make of the body. Moreover, the responses originating from body-sensing regions are likely to alter the operation of other perceptual systems, thus modulating not just the ongoing perception of the body but also that of the context in which body signaling is occurring. In the example of the wound, in parallel with a changed body, the ongoing cognitive processing will be altered as well. There is no way you will continue to enjoy whatever activity you were engaged in, as long as you experience the pain from the wound. This alteration of cognition is probably achieved by the release of molecules from brain stem and basal forebrain neuromodulator nuclei. Overall, these processes would lead to the assembly of qualitatively distinct maps, a contribution to the substrate of experiences of pain and pleasure.

Primordial Feelings

The issue of how perceptual maps of our body states become bodily feelings—how perceptual maps are *felt* and *experienced*—is not only central to the understanding of the conscious mind, it is integral to that understanding. One cannot fully explain subjectivity without knowing about the origin of feelings and acknowledging the existence of *primordial feelings,* spontaneous reflections of the state of the living body. In my view, primordial feelings result from nothing but the living body and precede any interaction between the machinery of life regulation and any object. Primordial feelings are based on the operation of upper-brain-stem nuclei, which are part and parcel of the life-regulation machinery. Primordial feelings are the primitives for all other feelings. I will return to this idea in Part III.

Mapping Body States and Simulating Body States

That the body, in most of its aspects, is continuously mapped in the brain and that a variable but considerable amount of the related information does enter the conscious mind is a proven fact. In order for the brain to coordinate physiological states in the body proper, which it can do without our being consciously aware of what is going on, the brain must be informed about the various physiological parameters at different regions of the body. The information must be current and consistent, from time to time, if it is to permit optimal control.

But this is not the only network that links body and brain. Around 1990 I proposed that in certain circumstances—for example, as an emotion unfolds—the brain rapidly constructs maps of the body comparable to those that would occur in the body had it actually been changed by that emotion. The construction can occur ahead of the emotional changes taking place in the body, or even *instead* of these changes. In other words, the brain can *simulate,* within somatosensing regions, certain body states, *as if* they were occurring; and because our perception of any body state is rooted in the body maps of the somatosensing regions, we perceive the body state as actually occurring even if it is not.[7]

At the time the "as-if body loop" hypothesis was first advanced, the evidence I could muster in its favor was circumstantial. It makes sense for the brain to know about the body state it is about to produce. The advantages of this sort of "advance simulation" are obvious from studies of the phenomenon of efference copy. Efference copy is what allows motor structures that are about to command the execution of a certain movement to inform visual structures of the likely consequence of that forthcoming movement in terms of spatial displacement. For example, when our eyes are about to move toward an object at the periphery of our vision, the visual region of the brain is forewarned of the impending movement and ready to smooth the transition to the new object without creating a blur. In other words, the visual region is allowed to anticipate the consequence of the movement.[8] Simulating a body state without actually producing it would reduce processing time and save energy. The as-if body loop hypothesis entails that the brain structures in charge of triggering a particular emotion be able to connect to the structures in which the body state corresponding to the emotion would be mapped. For example, the amygdala (a triggering site for fear) and the ventromedial prefrontal cortex (a triggering site for compassion) would have to connect to somatosensing regions, areas such as the insular cortex, SII, SI, and the somatosensory association cortices, where ongoing body states are continuously processed. Such connections exist, thereby rendering possible the implementation of the as-if body loop mechanism.

In recent years, more support for this hypothesis has come from several sources, one of which is a series of experiments by Giacomo Rizzolatti and his colleagues. In these experiments, which made use of electrodes implanted in the brains of monkeys, a monkey would watch an investigator perform a variety of actions. When a monkey saw the investigator move his hand, neurons in the monkey's brain regions related to its own hand movements became active, "as if" the monkey, rather than the investigator, were performing the action. But in reality the monkey remained immobile. The authors referred to the neurons that behaved in this manner as mirror neurons.[9]

So-called mirror neurons are, in effect, the ultimate as-if body device. The network in which those neurons are embedded achieves conceptually what I hypothesized as the as-if body loop system: the simulation, in the brain's body maps, of a body state that is not actually taking place in the organism. The fact that the body state simulated by mirror neurons is not the subject's body state amplifies the power of this functional resemblance. If a complex brain can simulate someone else's body state, one assumes that it would be able to simulate its own body states. A state that has already occurred in the organism should be easier to simulate since it has already been mapped by precisely the same somatosensing structures that are now responsible for simulating it. I suggest that the as-if system applied to others would not have developed had there not first been an as-if system applied to the brain's own organism.

The nature of the brain structures involved in the process reinforces the suggestive functional resemblance between the as-if body loop and the operation of mirror neurons. For the as-if body loop, I hypothesized that neurons in areas engaging emotion, such as the premotor-prefrontal cortex (in the case of compassion) and the amygdala (in the case of fear) would activate regions that normally map the state of the body and, in turn, move it to action. In humans such regions include the somatomotor complex in the Rolandic and parietal opercula as well as the insular cortex. All of these regions have a dual somatomotor role: they can hold a map of the body state, a sensory role, and they can participate in an action as well. By and large, this is what the neurophysiological experiments with monkeys uncovered. This is also consonant with human studies using magnetoencephalography[10] and functional neuroimaging.[11] Our own studies based on neurological lesions point in the same direction.[12]

Explanations of the existence of mirror neurons have emphasized the role that such neurons can play in allowing us to understand the actions of others by placing ourselves in a comparable body state. As we witness an action in another, our body-sensing brain adopts the body state we would assume were we ourselves moving, and it does so, in all probability, not by passive sensory patterns but by a preactivation of motor structures—ready for action but not allowed to act yet—and in some cases by actual motor activation.

How did such a complex physiological system evolve? I suspect that the system developed from an earlier as-if body loop system, which complex brains had long used to simulate their *own* body states. This would have had a clear and immediate advantage: rapid, energy-saving activation of the maps of certain body states, which were, in turn, associated with relevant past knowledge and cognitive strategies. Eventually the as-if system was applied to others and prevailed because of the equally obvious social advantages one could derive from knowing the body states of others, which are expressions of their mental states. In brief, I consider the as-if body loop system within each organism as the precursor to the operation of mirror neurons.

As we shall see in Part III, the fact that the body of a given organism can be represented in the brain is essential for the creation of the self. But the brain's representation of the body has another major implication: because we can depict our own body states, we can more easily simulate the equivalent body states of others. Subsequently, the connection we have established between our own body states and the significance they have acquired for us can be transferred to the simulated body states of others, at which point we can attribute a comparable significance to the simulation. The range of phenomena denoted by the word *empathy* owes a lot to this arrangement.

The Source of an Idea

I first gleaned the possibility described above many years ago in an odd and memorable episode. One summer afternoon when I was at work in the lab, I had gotten up from my chair and was walking across my office when I suddenly thought of my colleague B. I had no particular reason to think of him—I had not seen him recently, I did not need to talk to him, I had not read about him, I had no plans whatsoever to see him—and yet there he was present in my mind, the full recipient of my attention. One thinks of other people all the time, but this was different, because the presence was unexpected and demanded an explanation. Why was I thinking of Dr. B now?

Almost instantly a rapid succession of images told me what I needed to know. I mentally replayed my movements and realized that I had moved, for just a couple of moments, in a *manner* that was that of my colleague B. It had to do with the way I swung the arms and arched the legs. Now that I had discovered why I had been forced to think of him, I could picture his gait distinctly, in my mind's eye. But the fine point is that the visual images I had formed were prompted—better still, shaped—by the image of my own muscles and bones' adopting the distinctive motion patterns of my colleague B. In brief, I had just been walking *like* Dr. B; I had represented my animated skeletal frame in my own mind (technically, I had generated a somatosensory image); and finally I had recalled an appropriate visual counterpart for that particular musculoskeletal image, which turned out to be that of my colleague.

As the identity of the intruder was revealed, I also gleaned something intriguing about the human brain: I could adopt the characteristic motion of someone else by pure chance. (Or nearly so: in a further replay, I remembered I had seen B walking by my office window sometime earlier. I had processed him with little or no attention, largely nonconsciously.) I could transform the represented motion into a corresponding visual image, and I could recover from memory the identity of a person or persons that would fit the description. All of this was testimony to the close interconnections among an actual motion of the body, the representations of that motion in musculoskeletal and visual terms, and the memories that can be evoked in relation to some aspect of those representations.

This episode, enriched by additional observations and further reflection, made me realize how our connection to others occurs not just by visual images, language, and logical inference but also via something deeper in our flesh: the actions with which we can portray the movements of others. We can perform four-way translations among (1) actual movement, (2) somatosensory representations of movement, (3) visual representations of movement, and (4) memory. This episode would play a role in developing the notion of body simulation and its application in the as-if body loop.

Good actors, of course, use these devices in spades, knowingly or not. The manner in which some of the greats channel certain personalities into their compositions draws on this power to represent others, visually and auditorily, and then give them flesh in their own body. That is what inhabiting a role is all about, and when that process of transfer is decorated by unexpected, invented details, we get a performance of genius.

The Body-Minded Brain

The situation that emerges from the preceding facts and reflections is strange and unexpected but quite liberating.

We can all have our body in mind, at all times, providing us with a backdrop of feeling potentially available at every instant but noticeable only when it departs significantly from relatively balanced states and begins to register in the pleasantness or unpleasantness range. We have our body in mind because it helps govern behavior in all manner of situations that could threaten the integrity of the organism and compromise life. That particular function draws on the oldest kind of life regulation based on a brain. It harks back to simple body-to-brain signaling, to basic prompts for automated regulatory responses meant to assist with life management. But we simply have to marvel at what has been accomplished from such humble beginnings. Body mapping of the most refined order undergirds *both* the self process in conscious minds *and* the representations of the world external to the organism. The inner world has opened the way for our ability to *know* not only that very inner world but also the world around us.

The living body is the central locus. Life regulation is the need and the motivation. Brain mapping is the enabler, the engine that transforms plain life regulation into minded regulation and, eventually, into consciously minded regulation.

5

Emotions and Feelings

Situating Emotion and Feeling

In the quest to understand human behavior, many have tried to overlook emotion, but to no avail. Behavior and mind, conscious and not, and the brain that generates them, refuse to yield their secrets unless emotion (and the many phenomena that hide under its name) is factored in and given its due.

A discussion on the topic of emotion returns us to the matter of life and value. It requires a mention of reward and punishment, drives and motivations, and, of necessity, feelings. A discussion of emotions entails an investigation of the extremely varied devices of life regulation available in brains but inspired by principles and goals that anteceded brains and that, by and large, operate automatically and somewhat blindly, until they begin to be known to conscious minds in the form of feelings. Emotions are the dutiful executors and servants of the value principle, the most intelligent offspring yet of biological value. On the other hand, emotions' own offspring, the emotional feelings that color our entire life from cradle to grave, loom large over humanity by making certain that emotions are not ignored.

In Part III, when I will address the neural mechanisms behind the construction of the self, I will often invoke the phenomena of emotion and feeling because their machinery is used in the building of the self. The purpose of this chapter is to introduce that machinery briefly rather than present an overall review of emotions and feelings.

Defining Emotion *and* Feeling

Conversations on emotion face two major problems. One is the heterogeneity of phenomena that qualify for the label. As we have seen in Chapter 2, the value principle operates via reward and punishment devices as well as by drives and motivations, which are part and parcel of the emotion family. When we talk about emotions proper (say, fear, anger, sadness, or disgust), we talk, of necessity, of all those other devices too because they are constitutive components of each emotion and are independently involved in life regulation. Emotions proper are merely an integrated crown jewel of life regulation.

The other important problem is the distinction between emotion and feeling. Emotion and feeling, albeit part of a tightly bound cycle, are distinguishable processes. It makes no difference what words we choose to refer to these distinct processes, provided we acknowledge that the essence of emotion and the essence of feeling *are* different. Of course, there is nothing wrong with the words *emotion* and *feeling* to begin with, and they do perfectly well for the purpose, in English and in the many languages in which they have a direct translation. Let us begin, then, by defining those key terms in light of current neurobiology.

Emotions are complex, largely automated programs of *actions* concocted by evolution. The actions are complemented by a *cognitive* program that includes certain ideas and modes of cognition, but the world of emotions is largely one of actions carried out in our bodies, from facial expressions and postures to changes in viscera and internal milieu.

Feelings of emotion, on the other hand, are composite *perceptions* of what happens in our body and mind when we are emoting. As far as the body is concerned, feelings are images of actions rather than actions themselves; the world of feelings is one of perceptions executed in brain maps. But there is a qualification to be made here: the perceptions we call feelings of emotion contain a special ingredient that corresponds to the primordial feelings discussed earlier. Those feelings are based on the unique relationship between body and brain that privileges *interoception*. There are other aspects of the body being represented in emotional feelings, of course, but interoception dominates the process and is responsible for what we designate as the *felt* aspect of these perceptions.

The general distinction between emotion and feeling, then, is reasonably clear. While emotions are actions accompanied by ideas and certain modes of thinking, emotional feelings are mostly perceptions of what our bodies do during the emoting, along with perceptions of our state of mind during that same period of time. In simple organisms capable of behavior but without a mind process, emotions can be alive and well, but states of emotional feeling may not necessarily follow.

Emotions work when images processed in the brain call into action a number of emotion-triggering regions, for example, the amygdala or special regions of the frontal lobe cortex. Once any of these trigger regions is activated, certain consequences ensue—chemical molecules are secreted by endocrine glands and by subcortical nuclei and delivered to both the brain and the body (e.g., cortisol in the case of fear), certain actions are taken (e.g., fleeing or freezing; contraction of the gut, again in the case of fear), and certain expressions are assumed (e.g., a face and posture of terror). Importantly, in humans at least, certain ideas and plans also come to mind. For example, a negative emotion such as sadness leads to the recall of ideas about negative facts; a positive emotion does the opposite; the plans of action pictured in our minds are also in keeping with the overall signal of the emotion. Certain styles of mental processing are promptly instituted as an emotion develops. Sadness slows down thinking and may lead one to dwell on the situation that prompted it; joy may accelerate thinking and reduce attention to unrelated events. The aggregate of all these responses constitutes an "emotional state" unfolding in time, fairly rapidly, and then subsiding until new stimuli capable of causing emotions are introduced into the mind and begin yet another emotional chain reaction.

Feelings of emotion constitute the next step, following rapidly on the heels of emotion, the legitimate, consequential, ultimate achievement of the emotional process: the composite perception of all that has gone on during emotion—the actions, the ideas, the style with which ideas flow—fast or slow, stuck on an image, or rapidly trading one for another.

Seen from a neural perspective, the emotion-feeling cycle begins in the brain, with the perception and appraisal of a stimulus potentially capable of causing an emotion and the subsequent triggering of an emotion. The process then spreads elsewhere in the brain and in the body proper, building up the emotional state. In closing, the process returns to the brain for the feeling part of the cycle, although the return involves brain regions different from those in which it all started.

Emotion programs incorporate all the components of the life-regulation machinery that came along in the history of evolution, like the sensing and detection of conditions, the measurement of degrees of internal need, the incentive process with its reward and punishment aspects, the prediction devices. Drives and motivations are simpler constituents of emotion. This is why one's happiness or sadness alters the state of one's drives and motivations, immediately changing one's mix of appetites and desires.

Triggering and Executing Emotions

How are emotions triggered? Quite simply, by images of objects or events that are actually happening at the moment or that, having happened in the past, are now being recalled. The situation you are in makes a difference for the emotional apparatus. You may be actually inhabiting a scene of your life and responding to a musical performance or to the presence of a friend; or you may be alone and remembering a conversation that upset you the day before. Whether "live," reconstructed from memory, or created from scratch in one's imagination, the images initiate a chain of events. Signals from the processed images are made available to several regions of the brain. Some of those regions are involved in language, others in movement, others in manipulations that constitute reasoning. Activity in any of those regions leads to a variety of responses: words with which you can label a certain object; rapid evocations of other images that allow you to conclude something about an object; and so forth. Importantly, signals from the images that represent a certain object also land in regions capable of triggering specific kinds of emotional chain reaction. This is the case of the amygdala, for example, in situations of fear; or of the ventromedial prefrontal cortex in situations causing compassion. The signals are made available to all these sites. However, certain configurations of signals are likely to activate one particular site—provided the signals are sufficiently intense and the context is appropriate—and not activate the other sites where the same signals are also available. It is almost as if certain stimuli have the right key to open a certain lock, although the metaphor does not capture the dynamics and flexibility of the process. This is the case of fear-causing stimuli, which often activate the amygdalae and succeed in triggering the fear cascade. The same set of stimuli is not as likely to activate other sites. On occasion, however, certain stimuli are ambiguous enough to activate more than one site, leading to a composite emotional state. A bittersweet experience is the result, a "mixed" feeling arising from a mixed emotion.

In many respects, this is the strategy that the immune system uses to respond to invaders from outside the body. White blood cells called lymphocytes carry, on their surfaces, a huge repertoire of antibodies that match an equally large number of possible invader antigens. When one of these antigens enters the bloodstream and is allowed to make contact with lymphocytes, it will eventually bind with the antibody that best fits its shape. The antigen fits the antibody as a key does a lock, and the result is a reaction: the lymphocyte produces that antibody so abundantly that it will help destroy the invading antigen. I have proposed the term *emotionally competent stimulus* to echo the immune system and highlight the formal similarity of the emotional device to another basic device of life regulation.

What happens after "the key fits the lock" is nothing short of disturbing, in the proper sense of the term, since it amounts to an upset of the ongoing life state at multiple levels of the organism, from the brain itself to most divisions of the body proper. Again, in the case of fear, the upsets are as follows.

The nuclei in the amygdalae dispatch commands to the hypothalamus and to the brain stem that result in several parallel actions. The heart rate changes, and so do the blood pressure, the respiration pattern, and the state of contraction of the gut. The blood vessels in the skin contract. Cortisol is secreted into the blood, changing the metabolic profile of the organism in preparation for extra energy consumption. The muscles in the face move and adopt a characteristic mask of fear. Depending on the context in which the fear-causing images appear, one may then freeze in place or run away from the source of danger. Freezing or running, two very specific responses, are exquisitely controlled from separate regions of the brain stem's periaqueductal gray (PAG), and each response has its particular motor routine and physiological accompaniment. The freezing option automatically induces quiescence, shallow breathing, and a decrease in heart rate, which is an advantage in the attempt to be motionless and elude the attention of an attacker; the running option automatically increases heart rate and enhances blood circulation to the legs because one does need well-nourished leg muscles to run away. Moreover, if the brain selects the running option, the PAG automatically dampens the pain-processing pathways. Why? To better reduce the risk that a wound acquired on the run will paralyze the runner with intense pain.

The mechanism is so exquisite that yet another structure, the cerebellum, will struggle to modulate the expression of fear. This is why if one has been trained as a navy SEAL or as a marine, one's fear reaction will play out differently from that of someone who grew up as a potted plant.

Last, the processing of images in the cerebral cortex is itself affected by the ongoing emotion. For example, cognitive resources such as attention and working memory are adjusted accordingly. Certain topics of ideation are made unlikely—one is unlikely to think of sex or food when one runs away from a gunman.

Within a few hundred milliseconds, the emotional cascade manages to transform the state of several viscera, the internal milieu, the striated musculature of face and posture, the very pace of our mind, and themes of our thoughts. A disturbance indeed, as I am certain everyone will agree. When the emotion is strong enough, *upheaval,* the term used by the philosopher Martha Nussbaum, is an even better word.[1] All this effort, complicated in its orchestration and costly in the amount of energy it consumes—that is why being emotional is so bloody tiring—tends to have a useful purpose, and it often does. But it may not. Fear may be nothing but a false alarm induced by a culture gone awry. In those instances, rather than saving your life, fear is an agent of stress, and stress over time destroys life, mentally and physically. The upheaval has negative consequences.[2]

Some version of the entire collection of emotional changes in the body is conveyed to the brain via the mechanisms outlined in Chapter 4.

The Strange Case of William James

Before I turn to the physiology of feelings, I think it is appropriate to invoke William James and discuss the situation that his own words on the phenomena of emotion and feeling created, for himself and for emotion scholarship ever since.

A lapidary quote from James summarizes the issue, quickly and to the point.

Our natural way of thinking about these emotions is that the mental perception of some fact excites the mental affection called the emotion, and that this latter state of mind gives rise to the bodily expression. My thesis on the contrary is that the bodily changes follow directly the PERCEPTION of the exciting fact and that our feeling of the same changes as they occur IS the emotion.[3]

This is verbatim James, in 1884, including the capitalization of *perception* and *is*.

The importance of this idea cannot be overemphasized. James inverted the traditional sequence of events in the emotion process, and he interposed the body between the causative stimulus and the experience of emotion. There was no longer a "mental affection" called the emotion "giving rise to the body effects." There was, instead, the perception of a stimulus causing certain body effects. This was a bold proposal, and modern research entirely supports it. But the quote contains a major problem. After referring, in no uncertain terms, to "our feeling of the same changes," James confuses the issue by saying that the feeling, after all, "IS the emotion." This amounts to conflating emotion and feeling. James rejects emotion as a mental affection that causes body changes, only to accept emotion as a mental affection made of feelings of body changes, an entirely different arrangement from the one I presented earlier. It is unclear if this was unfortunate wording or an accurate expression of what James actually thought. Be that as it may, my view of emotions as action programs does not correspond to James's view as expressed in his text; his concept of feeling is not equal to mine. However, his idea of the mechanism for feeling is very much the same as my body loop mechanism of feeling. (James did not entertain an as-if mechanism, although a footnote in his text suggests that he saw the need for one.)

Most of the criticism that the James theory of emotion was to endure in the twentieth century was due to the wording of that paragraph. Leading physiologists such as Charles Sherrington and Walter

Cannon used James's words literally to conclude that their experimental data were incompatible with James's mechanism. Neither Sherrington nor Cannon was correct, but one cannot entirely fault them for their misprision.[4]

On the other hand, there are valid criticisms to be made of James's theory of emotion. For example, James left out stimulus appraisal altogether and confined the cognitive aspect of emotion to the perception of the stimulus and of body activity. For James, there was the perception of the exciting fact (which is equivalent to my emotionally competent stimulus), and the bodily changes followed directly. We know today that although things can actually happen this way, from fast perception to triggering of emotion, steps of appraisal tend to be interposed, a filtering and channeling of the stimulus as it makes its way through the brain and is led eventually to the trigger region. The appraisal stage can be very brief and nonconscious, but it needs to be acknowledged. James's view in this regard becomes a caricature: the stimulus always goes to the hot button and sets off the explosion. More important, the cognition generated by an emotional state is by no means confined to images of the stimulus and of the body changes, as James would have it. In humans, as we have seen, the emotion program also triggers certain cognitive changes that accompany the body changes. We can regard them as late components of the emotion or even as anticipated, relatively stereotyped components of the upcoming feeling of emotion. None of these reservations diminishes in any way James's extraordinary contribution.

Feelings of Emotion

Let me begin with a working definition. Feelings of emotion are composite perceptions of (1) a particular state of the body, during actual or simulated emotion, and (2) a state of altered cognitive resources and a deployment of certain mental scripts. In our minds, these perceptions are connected to the object that caused them.

Once it becomes clear that feelings of emotion are primarily perceptions of our body state during a state of emotion, it is reasonable to say that all feelings of emotion contain a variation on the theme of primordial feelings, whatever the primordial feelings of the moment are, augmented by other aspects of body change that may or not be related to interoception. It also becomes obvious that the substrate of such feelings in the brain should be found in the image-making regions of the brain, specifically in the somatosensing regions of two distinct sectors: the upper brain stem and the cerebral cortex. Feelings are states of mind based on a special substrate.

At the level of the cerebral cortex, the main region involved in feelings is the insular cortex, a sizable but quietly hidden part of the cerebral cortex located under both the frontal and parietal opercula. The insula, which does look like an island as the name implies, has several gyri. The front part of the insula is of old vintage, is related to taste and smell, and, just to confuse matters a bit, is a platform not only for feelings but also for the triggering of some emotions. It serves as a trigger point to a most important emotion: *disgust,* one of the oldest emotions in the repertoire. Disgust began its days as an automatic means of rejecting potentially toxic food and preventing it from entering the body. Humans can be disgusted not just by seeing spoiled food and the foul smell and taste that accompany it but by a variety of situations in which the purity of objects or behavior is compromised and there is "contamination." Importantly, humans are also disgusted by the perception of morally reprehensible actions. As a result, many of the actions in the human disgust program, including its typical facial expressions, have been co-opted by a social emotion: *contempt.* Contempt is often a metaphor for moral disgust.

The back part of the insula is made of modern neocortex, and the middle part is of intermediate phylogenetic age. The insular cortex has long been known to be associated with visceral function, representing the viscera and participating in their control. Along with the primary and secondary somatosensory cortices (known as SI and SII), the insula is a producer of body maps. Indeed, relative to the viscera and internal milieu, the insula is the equivalent of the primary visual or auditory cortices.

In the late 1980s I hypothesized a role for the somatosensory cortices in feelings, and I pointed to the insula as a likely provider of feelings. I wanted to move away from the hopeless idea of attributing the origin of feeling states to action-driving regions, such as the amygdalae. At the time, talking about emotion evoked sympathy if not derision, and suggesting a separate substrate for feelings evoked bewilderment.[5] Since 2000, however, we have known that activity in the insula is indeed an important correlate for every conceivable kind of feeling, from those that are associated with emotions to those that correspond to any shade of pleasure or pain, induced by a wide range of stimuli: hearing music one likes or hates; viewing pictures one loves, including erotic material, or pictures that cause disgust; drinking wine; having sex; being high on drugs; being low on drugs and experiencing withdrawal; and so forth.[6] The idea that the insular cortex is an important substrate of feelings is certainly correct.

But when it comes to the correlates of feelings, the insula is by no means the whole story. The anterior cingulate cortex tends to become active in parallel with the insula when we experience feelings. The insula and anterior cingulate are closely interlocked regions, the two being joined by mutual connections. The insula has dual sensory and motor functions, albeit biased toward the sensory side of the process, while the anterior cingulate operates as a motor structure.[7]

Most important, of course, is the fact that (as mentioned in the two previous chapters) several subcortical regions play a role in the construction of feeling states. At first glance, regions such as the nucleus tractus solitarius and the parabrachial nucleus have been seen as way stations for signals from the body's interior, as they convey them to a dedicated sector of the thalamus, which in turn signals to the insular cortex. But as indicated earlier, feelings likely begin to arise from activity in those nuclei, given their special status—they are the first recipients of information from the viscera and internal milieu with the ability to integrate signals from the entire range of the body's interior; in the upward progression from spinal cord to encephalon, these structures are the first capable of integrating and modulating signals about a comprehensive internal landscape—chest and abdomen, with their viscera—as well as visceral aspects of limbs and head.

To say that feelings arise subcortically is plausible given the evidence reviewed earlier: complete damage to the insular cortices in the presence of intact brain-stem structures is compatible with a wide range of feeling states; hydranencephalic children who lack insular and other somatosensory cortices but have intact brain-stem structures exhibit behaviors suggestive of feeling states.

No less important in the generation of feelings is a physiological arrangement that is central to my framework for mind and self: the fact that the brain regions involved in generating body maps and thus supporting feelings are part of a resonant loop with the very source of the signals they map. The upper brain-stem machinery in charge of body mapping interacts directly with the source of the maps it makes, in a tightly bound, near fusion of body and brain. Feelings of emotion emerge from a physiological system without parallel in the organism.

Let me conclude this section by recalling yet another important component of feeling states: all the thoughts prompted by the ongoing emotion. Some of those thoughts, as I noted earlier, are components of the emotion program, evoked as the emotion unfolds so that the cognitive context is in keeping with the emotion. Other thoughts, however, rather than being stereotypical components of the emotion program, are late cognitive reactions to the emotion under way. The images evoked by these reactions end up being a part of the feeling percept along with the representation of the object that caused the emotion in the first place, the cognitive component of the emotion program, and the perceptual readout of the body state.

How Do We Feel an Emotion?

In essence there are three ways of generating a feeling of emotion. The first and most obvious consists of having an emotion modify the body. Any emotion does this dutifully and swiftly because emotion *is* a program of action, and the result of the action is a change of body state.

Now, the brain is continuously generating a substrate for feelings because signals from the ongoing body state are continuously being reported, made use of, and transformed at the appropriate mapping sites. As an emotion unfolds, a specific set of changes occurs, and the *feeling of emotion maps* are the result of registering a *variation* superposed on the ongoing maps generated in the brain stem and in the insula. The maps constitute the substrate of a composite, multisite image.[8]

For the feeling state to be connected to the emotion, the causative object and the temporal relation between its appearance and the emotional response must be properly attended to. This is remarkably different from what happens in vision or hearing or smell. Because those other senses are focused on the world outside, the respective map-making regions can wipe their slates clean, as it were, and construct an infinity of patterns. Not so at the body-sensing sites, which are obligatorily turned to the inside and captive to what the body's infinite sameness feeds them. The body-minded brain is indeed a captive of the body and of its signaling.

The first way of generating feelings, then, requires what I call a body loop. But there are at least two other ways. One depends on the as-if body loop, introduced in Chapter 4. As the name suggests, it is a sleight of hand. The brain regions that initiate the typical emotion cascade can also command body-mapping regions, such as the insula, to adopt the pattern they would have adopted once the body signaled the emotional state to it. In other words, the triggering regions tell the insula to shape up, to configure its firing "as if" it were receiving signals describing emotional state X. The advantage of this bypass mechanism is obvious.

Since mounting a full-fledged emotional state takes a considerable amount of time and consumes a lot of precious energy, why not cut to the chase? No doubt this emerged in the brain precisely because of the economies of time and energy it introduced, and because smart brains are also extremely lazy. Anytime they can do less instead of more, they will, a minimalist philosophy they follow religiously.

There is only one hitch with the as-if mechanism. Like any other simulation, it is *not* quite like the real thing. I believe as-if feeling states are commonplace in all of us and certainly reduce the costs of our emotionality, but they are only attenuated versions of body-looped emotions. As-if patterns cannot possibly feel like the body-looped feeling states because they are simulations, not the genuine article, and also because it is probably more difficult for the weaker as-if patterns to compete with the ongoing body patterns than for the regular body loop versions to do so.

The other way of constructing feeling states consists of altering the transmission of body signals to the brain. As a consequence of natural analgesic actions or as a result of the administration of drugs that interfere with body signaling (painkillers, anesthetics), the brain receives a distorted view of what the body state really is at the moment. We know that in situations of fear in which the brain chooses the running option rather than freezing, the brain stem disengages part of the pain-transmission circuitry—a bit like pulling the phone plug. The periaqueductal gray, which controls these responses, can also command the secretion of natural opioids and achieve precisely what taking an analgesic would achieve: elimination of pain signals.

In the strict sense, we are dealing here with a *hallucination* of the body because what the brain registers in its maps and what the conscious mind feels do not correspond to the reality that might be perceived. Whenever we ingest molecules that have the power to modify the transmission or mapping of body signals, we play on this mechanism. Alcohol does it; so do analgesics and anesthetics, as well as countless drugs of abuse. It is patently clear that, other than out of curiosity, humans are drawn to such molecules because of their desire to generate feelings of well-being, feelings in which pain signals are obliterated and pleasure signals induced.

The Timing of Emotions and Feelings

In recent studies my colleague David Rudrauf has investigated the time course of emotions and feelings in the human brain using magnetoencephalography.[9] Magnetoencephalography is far less precise than functional magnetic resonance in terms of spatial localization of brain activity, but it offers a remarkable ability to estimate the time taken by certain processes in reasonably large sectors of the brain. We used this approach in these studies precisely because of the time feature.

Looking inside the brain, Rudrauf followed the time course of activity related to emotional and feeling reactions to pleasant or unpleasant visual stimuli. From the moment the stimuli were processed in the visual cortices to the moment the subjects first reported feelings, nearly five hundred milliseconds passed, or about half a second. Is this a little or a lot? It depends on the perspective. In "brain time" it is a huge interval, when one thinks that a neuron can fire in about five milliseconds. In "conscious mind time," however, it is not very much. It sits between the couple of hundred milliseconds we require to be conscious of a pattern in perception and the seven or eight hundred milliseconds we need to process a concept. Beyond the five-hundred-millisecond mark, however, feelings may linger for seconds or minutes, obviously reiterated in some sort of reverberation, especially if they are, well, big-time feelings.

The Varieties of Emotion

Attempts to describe the full range of human emotions or to classify them are not especially interesting. The criteria used for the traditional classifications are flawed, and any roster of emotions can be criticized for failing to include some and overincluding others. A vague rule of thumb suggests that we should reserve the term *emotion* for a reasonably complex program of actions (one that includes more than one or two reflexlike responses) triggered by an identifiable object or event, an emotionally competent stimulus. The so-called universal emotions (fear, anger, sadness, happiness, disgust, and surprise) are seen as meeting those criteria. Be that as it may, these emotions are certainly produced across cultures and are easily recognized because one part of their action program—their facial expressions—is quite characteristic. Such emotions are present even in cultures that lack distinctive names for the emotions. We owe to Charles Darwin the early recognition of this universality, not only in humans but in animals.

The universality of emotional expressions reveals the degree to which the emotional action program is unlearned and automated. At each performance, the emotion can be modulated, for example, with small changes of intensity or duration of component movements. The basic program routine, however, is stereotypical, at all the body levels at which it is executed—external motions; visceral changes in the heart, lungs, gut, and skin; and endocrine changes. The execution of the same emotion can vary from occasion to occasion but not enough to make it unrecognizable to the subject or to others. It varies as much as the interpretation of Gershwin's "Summertime" can change with different interpreters or even with the same interpreter on different occasions. It is still perfectly identifiable because the general contour of the behavior has been maintained.

The fact that emotions are unlearned, automated, and predictably stable action programs betrays their origin in natural selection and in the resulting genomic instructions. These instructions have been highly conserved across evolution and result in the brain's being assembled in a particular, dependable way, such that certain neuron circuits can process emotionally competent stimuli and lead emotion-triggering brain regions to construct a full-fledged emotional response. Emotions and their underlying phenomena are so essential for the maintenance of life and for subsequent maturation of the individual that they are reliably deployed early in development.

The fact that emotions are unlearned, automated, and set by the genome always raises the specter of genetic determinism. Is there nothing personable and educable about one's emotions? The answer is that there is plenty. The essential mechanism of the emotions in a normal brain is indeed quite similar across individuals, and a good thing too because it provides humanity, in diverse cultures, with a common ground of fundamental preferences on the matters of pain and pleasure. But while the mechanisms are distinctly similar, the circumstances in which certain stimuli have become emotionally competent for you are unlikely to be the same as for me. There are things that you fear that I do not, and vice versa; things you love and I do not, and vice versa; and many, many things that we both fear and love. In other words, emotional responses are considerably customized relative to the causative stimulus. In this regard, we are quite alike but not entirely. And there are other aspects to this individuation. Influenced by the culture in which we grew up, or as a result of individual education, we have the possibility of controlling, in part, our emotional expressions. We all know how public displays of laughter or crying are different across cultures and how they are shaped, even within membership in specific social classes. Emotional expressions resemble one another but are not equal. They can be modulated and made distinctly personal or suggestive of a social group.

The expression of emotions can doubtless be modulated voluntarily. But the degree of modulatory control of the emotions evidently cannot go beyond the external manifestations. Given that emotions include many other responses, several of which are internal and invisible to the naked eyes of others, the bulk of the emotional program is still executed, no matter how much willpower we apply to inhibit it. Most important, feelings of emotion, which result from the perception of the concert of emotional changes, still take place even when external emotional expressions are partially inhibited. Emotion and feeling have two faces, in keeping with their very different physiological mechanisms. When you encounter a stoic individual who stiffens his upper lip as tragic news arrives, do not surmise that he is not feeling anguish or fear. An old Portuguese adage captures this wisdom: "He who sees a face never gets to see the heart."[10]

Up and Down the Emotional Range

Besides the universal emotions, two commonly identified groups of emotion deserve special mention. Years ago I called attention to one of these groups and gave it a name: background emotions. Examples include *enthusiasm* and *discouragement,* two emotions that can be prompted by a variety of factual circumstances in one's life but also brought on by internal states such as disease and fatigue. Even more than with other emotions, the emotionally competent stimulus of background emotions may operate covertly, triggering an emotion without one's being aware of its presence. Reflection on a situation that has already happened, or consideration of a situation that is a mere possibility, can trigger such emotions. The resulting background feelings are just a small step up from primordial feelings. Background emotions are close relatives of moods but differ from them in their more circumscribed temporal profile and in the sharper identification of the stimulus.

The other major group of emotions is the social emotions. The label is a bit odd, since all emotions can be social and often are so, but the label is justifiable given the unequivocal social setting of these particular phenomena. Examples of the main social emotions easily justify the label: *compassion, embarrassment, shame, guilt, contempt, jealousy, envy, pride, admiration.* These emotions are indeed triggered in social situations, and they certainly play prominent roles in the life of social groups. The physiological operation of the social emotions is in no way different from that of other emotions. They require an emotionally competent stimulus;

they depend on specific triggering sites; they are constituted by elaborate action programs that involve the body; and they are perceived by the subject in the form of feelings. But there are some noteworthy differences. Most social emotions are of recent evolutionary vintage, and some may be exclusively human. This seems to be the case with admiration and with the variety of compassion that focuses on the mental and social pain of others rather than on physical pain. Many species, primates and the great apes in particular, exhibit forerunners of some social emotions. Compassion for physical predicaments, embarrassment, envy, and pride are good examples. Capuchin monkeys certainly appear to react to perceived injustices. Social emotions incorporate a number of moral principles and form a natural grounding for ethical systems.[11]

An Aside on Admiration and Compassion

The acts and objects we admire define the quality of a culture, as do our reactions to those who are responsible for those acts and objects. Without proper rewards, the admirable behaviors are less likely to be emulated. Likewise for compassion. Predicaments of every sort abound in daily life, and unless individuals behave compassionately toward those who face them, the prospects of a healthy society are greatly diminished. Compassion has to be rewarded if it is to be emulated.

What goes on in the brain when we feel admiration or compassion? Do the brain processes that correspond to such emotions and feelings resemble in any way those that we have identified for more basic emotions, such as fear, happiness, and sadness? Are they different? Social emotions seem so dependent on the environment in which one develops, so linked to educational factors, that they may seem a mere cognitive veneer applied lightly to the brain's surface. Also, it is important to examine how processing such emotions and feelings, which clearly involves the self of the beholder, engages, or does not, the brain structures that we have begun to associate with self states.

I set out to answer these questions with Hanna Damasio and with Mary Helen Immordino-Yang, whose consuming interest is the marriage of neuroscience and education and who was, for that very reason, attracted to this problem. We envisioned a study in which we would investigate, using functional magnetic resonance imaging, how stories can induce, in normal human beings, feelings of either admiration or compassion. We wanted to generate responses of admiration or compassion evoked by certain kinds of behavior, displayed in a narrative. We were not interested in having the experimental subjects recognize admiration or compassion when they witnessed them in someone else. We wanted the subjects to *experience* those emotions. From the beginning we knew that we wanted at least four distinct conditions, two for admiration, two for compassion. The admiration conditions were either admiration for virtuous acts (the admirable virtue of a great act of generosity) or admiration for acts of virtuosity (those of spectacular athletes or amazing musical soloists, for example). The compassion conditions, on the other hand, included compassion for physical pain (what one feels for the hapless victim of a street accident) and compassion for mental and social predicaments (what one feels for a person who lost his home in a fire, or lost her loved one to an incomprehensible disease).

The contrasts were very clear, all the more so once Mary Helen inventively assembled real stories along with an effective method to administer them to willing subjects in a functional imaging experiment.[12]

We tested three hypotheses. The first hypothesis had to do with the regions engaged by feeling admiration and compassion. The upshot of the experiment was unequivocal: the regions engaged were, by and large, the same as those engaged by the allegedly pedestrian basic emotions. The insula was alight in force, as was the anterior cingulate cortex, in all conditions. Upper-brain-stem regions were involved as well, as predicted.

This result certainly gave the lie to the idea that social emotions would not engage the machinery of life regulation to the same extent as their basic counterparts. The brain engagement ran deep, in keeping with the fact that our experiences of such emotions are deeply marked by body events. Jonathan Haidt's behavioral work on the processing of comparable social emotions reveals quite clearly how the body is engaged in such situations.[13]

The second hypothesis we tested concerned the central theme of this book: self and consciousness. We found that feeling these emotions engaged the posteromedial cortices (PMCs), a region we believe plays a role in constructing the self. This is in keeping with the fact that the subject's reaction to any of the stimulus stories required the person to become a full spectator and judge of the situation, a full empathizer with the protagonist's predicament, in the cases of compassion, and a potential prospective emulator of the protagonist's good deed, in the case of admiration.

We also found something we did not predict: the part of the PMCs that was most active in situations of admiration for skill and compassion for physical pain was quite distinct from the part of the PMCs that was most engaged by admiration for virtuous acts and compassion for mental pain. The split was striking, so much so that the PMC activity pattern related to one pair of emotions literally fit the PMC pattern related to the other, much like a missing piece in a puzzle.

The shared feature of one pair of conditions—skill and physical pain—was the involvement of the body in its external, action-oriented aspects. The shared feature of the other pair of conditions—the psychological pain of suffering and the virtuous act—was a mental state. The PMC result told us that the brain had recognized these shared features—physicality in one pair, mental states in the other—and paid them far more heed than the elementary contrast between admiration and compassion.

The likely explanation for this beautiful result comes from the different allegiances that the two parts of the PMC hold, in the brain of each subject, relative to the subject's own body. One sector relates closely to musculoskeletal aspects, the other to the very interior of the body, that is, to the internal milieu and viscera. The attentive reader will probably have guessed which goes with which. The physicality feature (skill, physical pain) goes with the musculoskeletal-related component. The mental feature (mental pain, virtue) goes with the internal milieu and viscera. Would you have it any other way?

There was one more hypothesis and one more result of note. We hypothesized that compassion for physical pain, being an evolutionarily older brain response—it is clearly present in several nonhuman species—should be processed faster by the brain than compassion for mental pain, something that requires the more complicated processing of a less immediately obvious predicament and that is likely to involve a wider compass of knowledge.

The results confirmed the hypothesis. Compassion for physical pain evokes faster responses in the insular cortex than does compassion for mental pain. The responses to physical pain not only rise faster but dissipate faster. The responses to mental pain take longer to establish themselves, but they also take longer to dissipate.

Despite the preliminary nature of this study, we have an initial glimpse of how the brain processes admiration and compassion. Predictably, the root of these processes runs deep in the brain and in the flesh. Also predictably, these processes are greatly affected by individual experience. All true, through and through, as it should be, for all emotions.

6

An Architecture for Memory

Somehow, Somewhere

"Will any of us ever see a train pulling out without hearing a few shots?" Dick Diver, the main character in Scott Fitzgerald's *Tender Is the Night,* asks his entourage as they wave good-bye to their friend Abe North in the Paris morning. Diver and company have just witnessed the unexpected: a desperate young woman has pulled a little pearl revolver from her purse and shot down her lover as the departing train whistled out of the Gare St. Lazare.

Diver's question is an evocative reminder of our brain's spectacular ability to learn composite information and reproduce it later, whether or not we wish it to, with considerable fidelity and from a variety of perspectives. Diver and company will forever come into train stations and *hear* imaginary shots in their minds, in a fainter but recognizable approximation of the sounds heard that morning, in an unwilled attempt to reproduce the auditory images experienced that morning. And because composite memories of events can be recalled from the representation of any of the parts that composed the event, they may also hear the shots when someone simply mentions departing trains, in any setting, not just when they see trains pulling out of stations, and they may also hear the shots when someone mentions Abe North (they were there because of him) or the Gare St. Lazare (that was where it occurred). This is also what happens to those who have been in a war zone and forever relive the sounds and sights of battle in haunting, unwanted flashbacks. Post-traumatic stress syndrome is the unwelcome side effect of an otherwise splendid ability.

It generally helps, as in this story, that the event to be remembered is emotionally salient, that it jitters the value scales. Provided that a scene has some value, provided that enough emotion was present at the time, the brain will learn multimedia sights, sounds, touches, feels, smells, and the like and will bring them back on cue. In time, the recall may grow faint. In time and with the imagination of a fabulist, the material may be embroidered upon, chopped to pieces, and recombined in a novel or screenplay. Step by step, what began as filmic nonverbal images may even morph into a fragmentary verbal account, remembered as much for the words in the tale as for the visual and auditory elements.

Now consider the marvel that is recall, and think of the resources the brain must possess to produce it. Beyond perceptual images in varied sensory domains, the brain must have a way of storing the respective patterns, somehow, somewhere, and must retain a *path* to retrieve the patterns, somehow, somewhere, for the attempted reproduction to work, somehow, somewhere. Once all of this happens and given the added gift of self, we *know* that we are in the middle of recalling something.

The ability to maneuver the complex world around us depends on this capacity to learn and recall—we recognize people and places only because we establish records of their likeness and bring some part of those records back at the right time. Our ability to imagine possible events also depends on learning and recall and is the foundation of reasoning and navigating the future and, more generally, for creating novel solutions for a problem. If we are to understand how all of this happens, we need to discover in the brain the secrets of the somehow and locate the somewhere. This is one of the intricate problems in contemporary neuroscience.

The approach to the problem of learning and recall depends on the level of operation we select to study. We have a growing understanding of what it takes, at the level of neurons and small circuits, for the brain to learn. For practical purposes, we know how synapses learn, and we even know, at the microcircuit level, some of the molecules and gene-expression mechanisms involved in learning.[1] We also know that specific parts of the brain play a main role in learning different kinds of information—objects such as faces, places, or words, on the one hand, and movements, on the other.[2] But many questions remain before the somehow and somewhere mechanisms can be fully elucidated. The purpose here is to outline a brain architecture that can further clarify the problem.

The Nature of Memory Records

The brain makes records of entities, the way they look and sound and act, and preserves them for later recall. It does the same for events. Usually the brain is assumed to be a passive recording medium, like film, onto which the characteristics of an object, as analyzed by sensory detectors, can be mapped faithfully. If the eye is the passive, innocent camera, the brain is the passive, virgin celluloid. This is pure fiction.

The organism (the body and its brain) interacts with objects, and the brain reacts to the interaction. Rather than making a record of an entity's structure, the brain actually *records the multiple consequences of the organism's interactions with the entity.* What we memorize of our encounter with a given object is not just its visual structure as mapped in optical images of the retina. The following are also needed: first, the sensorimotor patterns associated with viewing the object (such as eye and neck movements or whole-body movement, if applicable); second, the sensorimotor pattern associated with touching and manipulating the object (if applicable); third, the sensorimotor pattern resulting from the evocation of previously acquired memories pertinent to the object; fourth, the sensorimotor patterns related to the triggering of emotions and feelings relative to the object.

What we normally refer to as the memory of an object is *the composite memory of the sensory and motor activities related to the interaction between the organism and the object* during a certain period of time. The range of the sensorimotor activities varies with the value of the object and the circumstances, as does the retention of such activities. Our memories of certain objects are governed by our past knowledge of comparable objects or of situations similar to the one we are experiencing. Our memories are *prejudiced,* in the full sense of the term, by our past history and beliefs. Perfectly faithful memory is a myth, applicable only to trivial objects. The notion that the brain ever holds anything like an isolated "memory of the object" seems untenable. The brain holds a memory of what went on during an interaction, and the interaction importantly includes our own past, and often the past of our biological species and of our culture.

The fact that we perceive by engagement, rather than by passive receptivity, is the secret of the "Proustian effect" in memory, the reason why we often recall contexts rather than just isolated things. But it is also relevant to understanding how consciousness comes about.

Dispositions Came First, Maps Followed

The hallmark of brain maps is the relatively transparent connection between the thing represented—shape, movement, color, sound—and the map's contents. The pattern in the map has some patent correspondence to the thing it maps. In theory, if an intelligent observer could tumble onto the map in the course of her scientific wanderings, she would guess immediately what the map was supposed to stand for. We know this is not possible yet, although new imaging techniques are making good strides in that direction. In studies using functional magnetic resonance imaging (fMRI) in humans, multivariate pattern analyses demonstrate the presence of specific patterns of brain activity for certain objects seen or heard by the subject. In a recent study from our group (Meyer et al., 2010, cited in Chapter 3), we were able to detect patterns in the auditory cortex that correspond to what the subjects heard in their "mind's ear" (without any actual sound being heard). The results directly address the question posed by Dick Diver.

The biological development of mapping and its direct consequence—images and minds—is an insufficiently heralded transition in evolution. Transition from what? you may well ask. Transition from a mode of neural representation that had little overt connection to the thing represented. Let me give you an example. First, imagine that an object hits an organism, and an ensemble of neurons fires in response. The object might be pointed or blunt, large or small, handheld or self-propelled, made of plastic or steel or flesh. All that matters is that it *hits* the organism on some part of its surface, whereupon a neuron ensemble responds to the hit by becoming active, without actually representing the properties of the object. Now imagine another ensemble of neurons that fires upon receiving a signal from the first ensemble and then makes the organism move from its stationary position. Neither ensemble actually represented *where* the organism was in the first place, or *where* it should come to a stop, and neither ensemble represented the object's *physical properties*. What was needed was a detection of the hit, a command device, and the ability to move. That's all. What seems to have been represented by these brain ensembles is not maps but rather *dispositions,* know-how formulas that code for something like this: if hit from one side, move in the opposite direction for X number of seconds, regardless of the object hitting you or of where you are.

For a long, long time in evolution, brains operated on the basis of dispositions, and some of the organisms so equipped did perfectly fine in suitable environments. The dispositional network achieved a lot and got to be more and more complicated and wide-ranging in its achievement. But when the possibility of maps arose, organisms were able to go beyond formulaic responses and respond instead on the basis of the richer information now available in the maps. The quality of management improved accordingly. Responses became customized to objects and situations rather than being generic, and eventually the responses became more precise as well. Later, the dispositional, nonmapping networks would join forces with the networks that created maps, and as they did, organisms achieved an even greater management flexibility.

The fascinating fact, then, is that the brain did not discard its true and tried device (dispositions) in favor of the new invention (maps and their images). Nature kept both systems in operation and with a vengeance: it brought them together and made them work in synergy. As a result of the combination, the brain simply got richer, and that is the kind of brain we humans receive at birth.

Humans exhibit the most complicated example of that hybrid and synergic mode of operation, when we perceive the world, learn about it, recall what we have learned, and manipulate information creatively. We have inherited, from many prior species, abundant networks of dispositions that run our basic mechanisms of life management. They include the nuclei that control our endocrine system and the nuclei that serve the mechanisms of reward and punishment and the triggering and execution of the emotions. In a welcome novelty, these dispositional networks have been brought into contact with many systems of maps dedicated to imaging the world within and the world around. As a consequence, the basic mechanisms of life management influence the operation of the mapping regions in the cerebral cortex. But as I see it the novelty does not stop here, and the brains of mammals went one step further.

When human brains decided to create prodigiously large files of recorded images but lacked space to store them, they borrowed the disposition strategy to solve this engineering problem. They had their cake and ate it: they were able to fit numerous memories in a limited space but retain the ability to retrieve them rapidly and with considerable fidelity. We humans and our fellow mammals never had to microfilm various and sundry images and store them in hard-copy files; we simply stored a nimble formula for their reconstruction and used the existing perceptual machinery to reassemble them as best we could. We were always postmodern.

Memory at Work

Here is the problem, then. Besides creating mapped representations that result in perceptual images, the brain manages a no-less-remarkable achievement: it creates memory records of the sensory maps and plays back an approximation of their original content. This process is known as *recall*. Remembering a person or event or telling a story necessitates recall; recognizing objects and situations around us necessitates recall as well; so does thinking about objects with which we have interacted and about events we have perceived, and so does the entire imaginative process with which we plan for the future.

If we are to understand how memory works, we must understand how the brain establishes the record of a map as well as its location. Does it create a facsimile of the thing to be memorized, a sort of hard copy placed in a file? Or does it reduce the image to code—digitize it, as it were? Which? How? Where?

There is another critical *where* issue: Where is the record played back during recall, so that the essential properties of the original image can be recuperated? When Dick Diver, in *Tender Is the Night,* comes to hear the shots again, where in his brain are they being played back? When you think of a friend you lost or of a house you lived in, you conjure up a collection of images of those entities. They are less vivid than the real thing or a photograph. But recalled images can maintain the basic properties of the original, so much so that an ingenious cognitive neuroscientist, Steve Kosslyn, has been able to estimate the relative size of an object recalled and inspected in mind.[3] Where are the images reconstructed so that we can study them in our reverie?

The traditional answers (although assumptions would be a better word) to this question get their inspiration from a conventional account of sensory perception. Accordingly, different early sensory cortices (largely in the back sections of the brain) bring forward the components of perceptual information by brain pathways to so-called multimodal cortices (largely in the front sections), which integrate them. Perception would operate on the basis of a cascade of processors going in one direction. The cascade would extract, step by step, more and more refined signals, first in the sensory cortices of a single modality (e.g., visual) and later in multimodal cortices, those that receive signals from more than one modality (e.g., visual, auditory, and somatic). The cascade would follow, in general, a caudo-rostral direction (back to front) and would culminate in the anterior temporal and frontal cortices, where the most integrated representations of the ongoing multisensory apprehension of reality are presumed to occur.

These assumptions are captured by the notion of a "grandmother cell." A grandmother cell is a neuron somewhere near the top of the processing cascade (e.g., the anterior temporal lobe) whose activity would, in and of itself, comprehensively represent our grandmother when we *perceive* her. Such single cells (or small ensembles of cells) would hold an all-encompassing representation of objects and events during perception. Not only that, they would also hold a *record* of those perceived contents. The memory records would be where the grandmother cells are. Even more grandly, and in direct response to the question posed earlier, reactivated grandmother cells would allow the playback of those same perceived contents in their entirety, right there and then. In brief, activity in those neurons would account for the recall of varied and properly integrated images, the face of your grandmother or Dick Diver's train station shots included. That would be the *where* of recall.

I regard the above account as unlikely. By this account, damage to the up-front temporal and frontal lobe cortices, the anterior brain regions, should preclude *both* normal perception *and* normal recall. Normal perception would collapse because the neurons capable of creating the fully integrated representation of a cohesive perceptual experience would no longer be functional. Normal recall would collapse because the same cells that support integrated perception also support integrated memory records.

Alas for the health of the traditional view, this prediction is not borne out by the reality of neuropsychological findings. The highlights of this dissenting reality are as follows. Patients with damage to the anterior brain regions—frontal and temporal—report normal perception and display only selective deficits in the recall and recognition of unique objects and events.

The patients may describe in great detail the contents of a picture they are shown, describe the picture correctly, as being that of a party (birthday, wedding), and yet fail to recognize that it was their own party. Anterior damage compromises neither the integrated perception of the whole scene nor the interpretation of its meaning. Nor does it compromise perception of the numerous objects that comprise the picture and the retrieval of their meaning—people, chairs, tables, birthday cake, candles, festive attire, and so forth. Anterior damage permits the integrated view and the view of the parts. It takes an entirely different placement of damage to compromise the access to the separable memory components, those that correspond to varied objects or to features of objects, such as color or movement. Such access is compromised, but only by damage involving sectors of the cerebral cortex positioned farther back in the brain, near the main sensory and motor regions.

In conclusion, damage to the integrative, associative cortices does not preclude integrated perception, or recall of the parts that constitute a set, or recall of the meaning of nonunique sets of objects and features. Such damage makes one specific and major dent in the recall process: *it precludes the recall of uniqueness and specificity of objects and scenes*. A unique birthday party continues to be a birthday party, but it is no longer someone's *specific* birthday party, complete with place and date line. Only damage to the mind-making early sensory cortices and their surrounds precludes recall of the information that once was processed by those cortices and recorded nearby.

A Brief Aside on Kinds of Memory

The distinctions we can make among different types of memory relate not only to the subject matter that is the focus of recall, but also to the range of circumstances surrounding that focus, as represented in a particular recall situation. In this light, several traditional labels commonly applied to memories (*generic* versus *unique, semantic* versus *episodic*) do not capture the wealth of the phenomenon. For instance, if I am asked about a particular house where I once lived, either through a verbal prompt or through a photograph, I am likely to recall a wealth of memoranda related to my personal experiences of that house; this includes the reconstruction of sensorimotor patterns of varied modality and type, such that even personal feelings may be reenacted. If, instead, I am asked to evoke the general concept of house, I may well recall the same unique house, in my mind's eye, and then go about articulating the generic concept of house. In those circumstances, however, the nature of the question alters the course of the recollection process. The purpose of the second request probably inhibits the evocation of the rich personal details that were so prominent in the previous one. Rather than a personal remembrance, I will simply process a set of facts that satisfy my need of the moment, which is to define *house*.

The distinction between the first and the second examples resides with the degree of complexity in the recollection process. That complexity can be measured by the number and variety of items recalled in connection with a particular target or event. In other words, *the larger the sensorimotor context that is reenacted relative to a particular entity or event, the greater the complexity*. The memory of unique entities and events, namely, those that are both unique and personal, requires high-complexity contexts. We can glean a hierarchical progression of complexity here: unique-personal entities and events require the highest complexity; unique-nonpersonal entities and events are next; nonunique entities and events require least.

For practical purposes, it is useful to say that a given term is recalled at one of the above levels—nonunique or unique-personal, say. That distinction is roughly comparable to *the semantic/episodic* distinction, or the *generic/contextual* distinction.

It is also useful to preserve the distinction between *factual* memory and *procedural* memory because it does capture a fundamental divide between "things"—entities that have a certain structure, in repose—and the "movement" of things in space and in time. Even here, however, the distinction can get dicey.

In the end, the validity of these categories of memory resides with whether the brain honors the distinction. By and large, the brain honors distinctions between unique and nonunique levels of processing at the level of recall, and between factual and procedural kinds of memory, both in the making of a memory and in the recall.

A Possible Solution to the Problem

Reflection on these observations led me to propose a model of neural architecture aimed at accounting for recall and recognition.[4] What the model accomplished is as follows.

Images can be experienced during perception and during recall. It would be impossible to store the maps that underlie all images one has experienced, in their original format. For example, the early sensory cortices are continuously constructing maps about the current environment and have no resources to store discarded maps. But in brains such as ours, thanks to the reciprocal connections between the map-making brain space and the dispositional space, maps can be recorded in dispositional form. In such brains, dispositions are also a space-saving mechanism for information storage. Finally, dispositions can be used to reconstruct the maps in early sensory cortices, in the format in which they were first experienced.

The model took into account the neuropsychological findings described earlier and posited that the cell ensembles at the top levels of the processing hierarchies would not hold explicit representations of the maps for objects and events. Rather, the ensembles would hold know-how, that is, *dispositions,* for the eventual reconstruction of explicit representations when they become needed. In other words, I was using the simple disposition device that I introduced earlier, but this time, rather than commanding a trivial movement, the disposition was *commanding the process of reactivating and putting together aspects of past perception,* wherever they had been processed and then locally recorded. Specifically, the dispositions would act on a host of early sensory cortices originally engaged by perception. The dispositions would do so by dint of connections diverging from the disposition site back to early sensory cortices. In the end, the locus where memory records would actually be played back would not be that different from the locus of original perception.

Convergence-Divergence Zones

The main piece of the proposed framework was a neural architecture of cortical connections that had convergent and divergent signaling properties relative to certain nodes. I called the nodes *convergence-divergence zones* (CDZs). CDZs recorded the *coincidence* of activity in neurons hailing from different brain sites, neurons that had been made active by, for example, the mapping of a certain object. No part of the overall map of the object had to be permanently re-represented in the CDZs, to be placed in memory. Only the coincidence of signals from neurons linked to the map needed to be recorded. To reconstitute the original map and thus produce recall, I proposed the mechanism of *time-locked retroactivation.* The term *retroactivation* pointed to the fact that the mechanism required a process of "going back" in order to induce activity; *time-locked* called attention to another requirement: it was necessary to retroactivate the components of a map roughly within the same time interval, so that what occurred simultaneously (or nearly so) in perception could be reinstated simultaneously (or nearly so) in recall.

The other critical element in the framework consisted of positing a division of labor between two kinds of brain systems, one that managed maps/images and another that managed dispositions. As far as the cerebral cortices were concerned, I proposed that the *image space* consisted of several islands or early sensory cortices—for example, the ensemble of visual cortices that encircle the primary visual cortex (area 17 or V_1), the ensemble of auditory cortices, that of somatosensory cortices, and so forth.

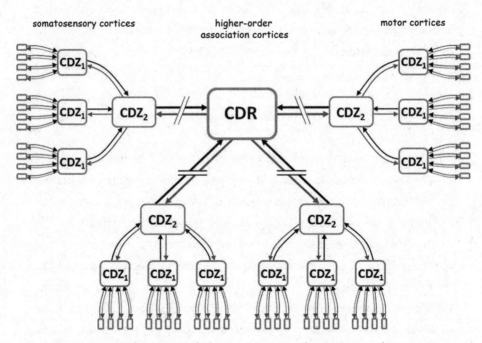

Figure 6.1: Schematics of the convergence-divergence architecture. Four hierarchical levels are depicted. The primary cortical level is shown in small rectangular boxes, and three levels of convergence-divergence (larger boxes) are marked CDZ_I, CDZ_2, and CDR. Between CDZ levels and CDR levels (interrupted arrows), numerous intermediate CDZs are possible. Note that, throughout the network, every forward projection is reciprocated by a return projection (arrows).

The cortical *dispositional space* included all the higher-order association cortices in temporal, parietal, and frontal regions; in addition, an old set of dispositional devices remained beneath the cerebral cortex in the basal forebrain, basal ganglia, thalamus, hypothalamus, and brain stem.

In brief, the image space is the space where explicit images of all sensory types occur, including both the images that become conscious and those that remain unconscious. The image space is located in the map-making brain, the large territory formed by the aggregate of all the early sensory cortices, the regions of cerebral cortex located in and around the entry point of visual, auditory, and other sensory signals into the brain. It also includes the territories of the nucleus tractus solitarius, parabrachial nucleus, and superior colliculi, which have image-making capability.

The *dispositional space* is that in which dispositions hold the knowledge base as well as the devices for the reconstruction of that knowledge in recall. It is the source of images in the process of imagination and reasoning and is also used to generate movement. It is located in the cerebral cortices that are not otherwise occupied by the image space (the higher-order cortices and parts of the limbic cortices) and in numerous subcortical nuclei. When dispositional circuits are activated, they signal to other circuits and cause images or actions to be generated.

The contents exhibited in the image space are *explicit,* while the contents of the dispositional space are *implicit*. We can access the contents of images, if we are conscious, but we never access the contents of dispositions directly. Of necessity, *the contents of dispositions are always unconscious*. They exist in encrypted and dormant form.

Dispositions produce a variety of results. At a basic level, they can generate actions of many kinds and many levels of complexity—the release of a hormone into the bloodstream; the contraction of muscles in viscera or of muscles in a limb or in the vocal apparatus. But cortical dispositions also hold records of an image that was actually perceived on some previous occasion, and they participate in the attempt to reconstruct a sketch of that image from memory. Dispositions also assist with the processing of a currently perceived image, for instance, by influencing the degree of attention accorded to the current image. We are never aware of the knowledge necessary to perform any of these tasks, nor are we ever aware of the intermediate steps that are taken. We are aware only of results, like a state of well-being, the racing of the heart, the movement of a hand, the fragment of a recalled sound, the edited version of the ongoing perception of a landscape.

Our memories of things, of properties of things, of people and places, of events and relationships, of skills, of life-management processes—in short all of our memories, inherited from evolution and available at birth or acquired through learning thereafter—exist in our brains in dispositional form, waiting to become explicit images or actions. *Our knowledge base is implicit, encrypted, and unconscious.*

Dispositions are not words; they are abstract records of potentialities. The basis for the enactment of words or signs also exists as dispositions before they come to life in the form of images and actions, as in the production of speech or sign language. The rules with which we put words and signs together, the grammar of a language, are also held as dispositions.

More on Convergence-Divergence Zones

A convergence-divergence zone (CDZ) is an ensemble of neurons within which many feedforward-feedback loops make contact. A CDZ receives "feedforward" connections from sensory areas located "earlier" in the signal-processing chains, which begin at the entry point of sensory signals in the cerebral cortex. A CDZ sends reciprocal feedback projections to those originating areas. A CDZ also sends "feedforward" projections to regions located in the next connectional level of the chain and receives return projections from them.

CDZs are microscopic and are located within convergence-divergence regions (CDRegions), which are macroscopic. I envision the number of CDZs to be on the order of many thousands. On the other hand, CDRegions number in the dozens. CDZs are micronodes; CDRegions are macronodes.

CDRegions are located at strategic areas in association cortices, areas toward which several major pathways converge. You can visualize CDRegions as hubs on an airline map. Think of Chicago, Washington, D.C., New York, Los Angeles, San Francisco, Denver, or Atlanta. Hubs receives airplanes along the spokes that come into the hub, and they return airplanes back along the same spokes. Importantly, hubs themselves are interconnected, though some are more peripheral than others. Finally, some hubs are bigger than others, which simply means that more CDZs are living under their umbrellas.

We know from experimental neuroanatomical studies that such patterns of connectivity exist in the primate brain.[5] We also know from recent magnetic resonance neuroimaging studies using diffusion spectrum techniques that such patterns exist in humans.[6] We shall see, in the chapters ahead, that CDRegions play an important role in producing and organizing critical contents of the conscious mind, including those that make up the autobiographical self.

Both CDRegions and CDZs come into existence under genetic control. As the organism interacts with the environment during development, synaptic strengthening or weakening modifies convergence regions significantly and massively modifies CDZs. Synaptic strengthening occurs when external circumstances match the survival needs of the organism.

In brief, the job I envision for CDZs consists of re-creating separate sets of neural activity that were once approximately simultaneous during perception—that is, that coincided during the time window necessary for us to attend to them and be conscious of them. To achieve this, the CDZ would prompt an extremely fast sequence of activations that would make separate neural regions come online in some order, the sequence being imperceptible to consciousness.

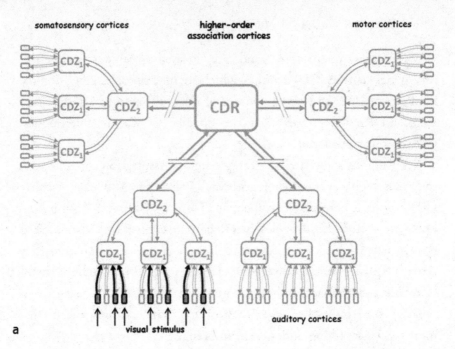

a

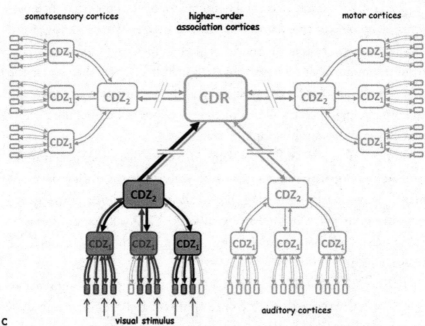

c

Figure 6.2: Using the CD architecture to recall memories prompted by a specific visual stimulus. In panels *a* and *b*, a certain incoming visual stimulus (selective set of small filled-in boxes) prompts forward activity in CDZs of levels 1 and 2 (bold arrows and filled-in boxes). In panel *c*, forward activity activates specific CDRs, and in panel *d*, retroactivation

In this architecture, knowledge retrieval would be based on relatively simultaneous, attended activity in many early cortical regions, engendered over several reiterations of such reactivation cycles. Those separate activities would be the basis of reconstructed representations. The level at which knowledge is retrieved would depend on the scope of multiregional activation. In turn, this would depend on the level of CDZ that is activated.[7]

The Model at Work

What evidence is there that the convergence-divergence model fits reality? Recently, my colleague Kaspar Meyer and I reviewed a large number of studies in the areas of perception, imagery, and mirror processing and considered the results from the perspective of the convergence-divergence model.[8] Many of the results we reviewed constitute interesting tests of the model. Here is a case in point.

In a conversation with another person, we hear the speaker's voice and see the speaker's lips move at the same time. The CDZ model predicts that, as a certain lip movement repeatedly occurs along with its specific sound counterpart, the two neural events, in the early visual and auditory cortices, respectively, become associated in a shared CDZ. In the future, when we are confronted with only one part of that scene—for example, as we watch a specific lip movement in a muted video clip—the activity pattern induced in the early visual cortices will trigger the shared CDZ, and the CDZ will retroactivate, in the early auditory cortices, the representation of the sound that originally accompanied the lip movement.

In keeping with the CDZ framework, reading lips in the absence of any sound induces activity in the auditory cortices, and the evoked activity patterns overlap with those elicited during the perception of spoken words.[9] The auditory map of the sound becomes an integral part of the representation of the lip movement. The CDZ framework

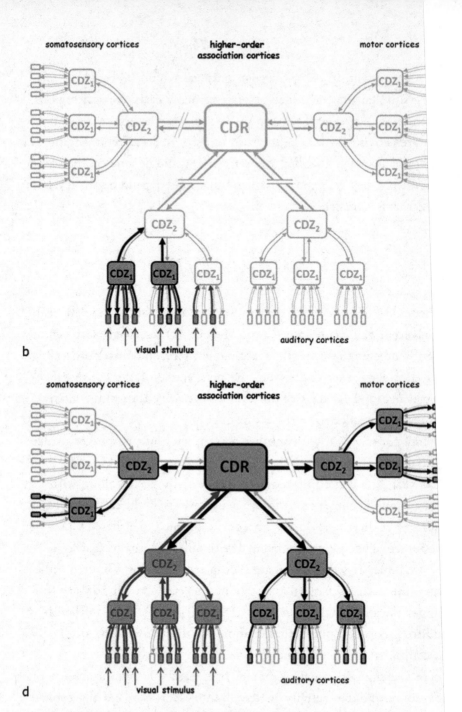

from CDRs prompts activity in early somatosensory, auditory, motor, and other visual cortices (bold arrows, filled-in boxes). Retroactivation generates displays in "image space" as well as movement (selective set of small filled-in boxes).

explains how one can hear sound, in the mind's eye, upon receiving the appropriate visual stimulus, or vice versa.

Should anyone regard the brain's feat of synchronizing visuals and sound as a trivial achievement, just think of the discomfort and irritation one feels when the quality of a film projection fails and the sound track and visual image go out of sync. Or worse yet, when one has to watch a great Italian film badly dubbed in unsynced English. A variety of other perceptual studies involving other sensory modalities (smell, touch) and even neuropsychological studies in nonhuman primates yield results that are satisfactorily explained by the CDZ model.[10]

Another interesting set of data comes from studies of mental imagery. The process of imagination, as the term suggests, consists of the recall of images and their subsequent manipulation—cutting, enlarging, reordering, and so forth. When we use our imagination, does imagery take place in the form of "pictures" (visual, auditory, and so forth), or does it rely on mental descriptions resembling those of language?[11] The CDZ framework supports the picture account. It proposes that comparable regions are activated when objects or events are perceived and when they are recalled from memory. The images constructed during perception are *re*-constructed during the process of imagery. They are approximations rather than replicas, attempts at getting back at past reality and thus not quite as vivid or accurate.

A large number of studies indicates unequivocally that imagery tasks in modalities such as visual and auditory usually evoke brain activity patterns that overlap to a considerable extent with the patterns observed during actual perception,[12] while the results from lesion studies also provide compelling evidence for the CDZ model and the pictorial account of imagination. Focal brain damage often causes simultaneous deficits in perception and imagery. An example is the inability to both perceive and imagine colors caused by damage to the occipitotemporal region. Patients with focal damage to this region see their visual world in black and white, literally in shades of gray. The patients are unable to "imagine" color in their minds. They know perfectly well that blood is red, and yet they cannot picture red in their mind's eye, any more that they can see red when they look at a red-colored chip.

Evidence from both functional imaging and lesion studies suggests that recall of objects and events relies, at least in part, on activity near the points where sensory signals enter the cortex, as well as near motor output sites. It is certainly no coincidence that these are the sites engaged in the original perception of objects and events.

Mirror neuron research also provides evidence that a convergence-divergence architecture is a satisfactory means to explain certain complex behaviors and mental operations. The key finding in mirror neuron research (Chapter 4) is that the mere observation of an action leads to activity in motor-related areas.[13] The CDZ model is ideal for explaining this observation. Consider what happens when we act. An action does not consist merely of a sequence of movements generated by the brain's motor regions. The action encompasses simultaneous sensory representations that arise in the somatosensory, visual, and auditory cortices. The CDZ model suggests that the repeated co-occurrence of the varied sensorimotor maps that describe a specific action leads to repeated convergent signals toward a particular CDZ. At a later occasion, when the same action is perceived, say visually, the activity generated in visual cortices activates the pertinent CDZ. Subsequently, the CDZ uses divergent back projections toward early sensory cortices to reactivate the related associations of the action in modalities such as somatosensory and auditory. The CDZ can also signal toward motor cortices and generate a mirror movement. From our perspective, mirror neurons are CDZ neurons involved in movement.[14]

According to the CDZ model, mirror neurons alone would not enable observers to grasp the meaning of an action. CDZs do not hold the meaning of objects and events themselves; they reconstruct meaning via time-locked multiregional retroactivation into varied early cortices. Since mirror neurons are likely to be CDZs, the meaning of an action cannot be subsumed by mirror neurons only. A reconstruction of varied sensory maps previously associated with the action needs to be carried out under the control of the CDZs in which a linkage to those original maps has been recorded.[15]

The How and Where of Perception and Recall

The perception or recall of most objects and events depends on activity in varied image-making regions of the brain and often involves parts of the brain related to movement as well. This highly dispersed pattern of activity occurs within the *image space*. It is this activity, rather than the activity to be found in neurons at the front end of the processing chains, that allows us to perceive explicit images of objects and events. From a functional as well as an anatomical standpoint, the activity at the end of the processing chains occurs within *dispositional space*. The dispositional space is made up of CDZs and CDRs, in association cortices, which are not image-making cortices. The dispositional space guides the image-making but is not involved in displaying images itself.

In this sense, the dispositional space contains "grandmother cells," defined liberally as neurons whose activity correlates with the presence of a specific object, but not as neurons whose activity permits, in and of themselves, explicit mental images of objects and events. Neurons in anterior medial temporal cortices can indeed respond to unique objects, in perception or recall, with high specificity, suggesting that they receive convergent signals.[16] But the mere activation of those neurons, without the retroactivation that would follow from it, would not allow us to recognize our grandmother or remember her. To recognize or remember our grandmother, we must reinstate a substantial part of the collection of explicit maps that, in their entirety, represent her meaning. Like mirror neurons, so-called grandmother neurons are CDZs.

A

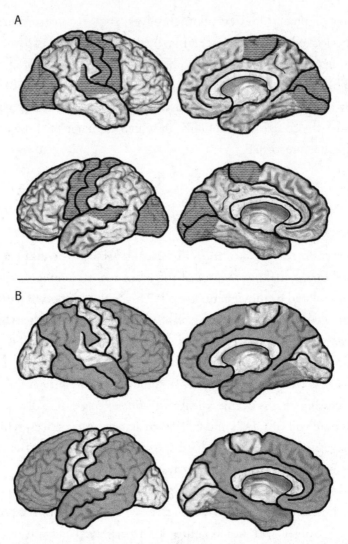

B

Figure 6.3: The image space (mapped) and the dispositional space (nonmapped) in the cerebral cortex. The image space is depicted in the shaded areas of the four A panels, along with the primary motor cortex.

The dispositional space is depicted in the four B panels, again marked by shading.

The separate components of the image space resemble islands in the ocean of dispositional space shown in the four bottom panels.

They enable the time-locked multiregional retroactivation of explicit maps in early sensorimotor cortices.

In conclusion, the CDZ framework posits two somewhat separate "brain spaces." One space constructs explicit maps of objects and events during perception and reconstructs them during recall. In both percept and recall, there is a manifest correspondence between the properties of the object and the map. The other space holds dispositions rather than maps, that is, implicit formulas for how to reconstruct maps in the image space.

The explicit image space is constituted by the aggregate of early sensorimotor cortices. When I talk about "workspace" in relation to the sites where images are assembled, I think of such a space, as a playground for the puppetry we behold in the conscious mind. The implicit, dispositional space is constituted by the aggregate of association cortices. This is the space where many unwitting puppet masters pull the invisible puppet strings.

The two spaces point to different ages in brain evolution, one in which dispositions sufficed to guide adequate behavior and another in which maps gave rise to images and to an upgrade of the quality of behavior. Today they are seamlessly integrated.

PART III

Being Conscious

7

Consciousness Observed

Defining Consciousness

Open a standard dictionary in search of a definition of *consciousness,* and you are likely to find some variation of the following: "consciousness is a state of awareness of self and surroundings." Substitute *knowledge* for *awareness,* and *own existence* for *self,* and the result is a statement that does capture some essential aspects of consciousness as I see it: consciousness is *a state of mind in which there is knowledge of one's own existence and of the existence of surroundings.* Consciousness is *a state of mind*—if there is no mind there is no consciousness; consciousness is a *particular* state of mind, enriched by a sense of the particular organism in which a mind is operating; and the state of mind includes knowledge to the effect that the said existence is *situated,* that there are objects and events surrounding it. Consciousness is a state of mind with a self process added to it.

The conscious state of mind is experienced in the exclusive, first-person perspective of each of our organisms, never observable by anyone else. The experience is owned by each of our organisms and by no other. But even though the experience is exclusively private, we can still adopt a relatively "objective" view toward it. For example, I adopt such a view in the attempt to glean a neural basis for the self-as-object, the material me. An enriched material me is also capable of delivering

knowledge to the mind. In other words, the self-as-object can also operate as knower.

We can amplify this definition by saying that conscious mind states
always have content (they are always about something) and that some of
the contents tend to be perceived as integrated collections of parts (as,
for example, when we both see and hear a person speaking and walking
toward us); by saying that conscious states of mind reveal distinct qualitative properties relative to the different contents one comes to know (it
is qualitatively different to see or listen, to touch or taste); and by saying
that conscious states of mind contain an obligate aspect of *feeling*—they
feel like something to us. Finally, our provisional definition must say
that conscious states of mind are possible only when we are awake,
although a partial exception to this definition applies to the paradoxical
form of consciousness that occurs during sleep, in dreaming. In conclusion, in its standard form, consciousness is a state of mind that occurs
when we are awake and in which there is private and personal knowledge of our own existence, situated relative to whatever its surround
may be at a given moment. Of necessity, conscious states of mind handle knowledge based on different sensory material—bodily, visual,
auditory, and so forth—and manifest varied qualitative properties for
the different sensory streams. Conscious states of mind are *felt*.

When I talk about consciousness, I am not referring simply to wakefulness, a common misuse that comes from the fact that when wakefulness is lost, consciousness is often lost as well. (I will address these issues
in the pages ahead.) The definition also makes clear that the term *consciousness* does *not* refer simply to a plain mind process, without the self
feature. Unfortunately, taking consciousness as mere mind is a common
use of the term—a misuse, I think. People often refer to "something
being in consciousness" to mean that something is "in mind" or that
something has become a prominent content of mind, as in "the issue of
global warming has finally penetrated the consciousness of Western
nations"; a significant number of contemporary consciousness studies
treat consciousness as mind. Nor does consciousness, as used in this

book, stand for "self-consciousness" as meant in "John got more and more self-conscious as she continued to stare at him"; or "conscience," a complex function that does require consciousness but goes well beyond it and pertains to moral responsibility. Finally, the definition does not refer to consciousness as in the colloquial sense of James's "stream of consciousness." The phrase is often meant to signify the plain contents of mind as they flow forward in time, like water in a riverbed, rather than the fact that such contents incorporate subtle or not-so-subtle aspects of subjectivity. References to consciousness in the context of Shakespeare's soliloquies or Joyce's often use this simpler view of consciousness. But the original authors were obviously exploring the phenomenon in its full sense, writing from the perspective of a character's self, so much so that Harold Bloom has suggested that Shakespeare may have single-handedly introduced the phenomenon of consciousness into literature. (But see James Wood's alternative and entirely plausible claim that consciousness did enter literature by way of the soliloquy but far earlier—in prayer, for example, and in Greek tragedy.)[1]

Breaking Consciousness Apart

Consciousness and wakefulness are not the same thing. Being awake is a prerequisite of being conscious. Whether one falls asleep naturally or is forced to sleep by anesthesia, consciousness vanishes in its standard format, the only partial exception being the particular conscious state that accompanies dreams, and that in no way contradicts the wakefulness prerequisite because dream consciousness is not standard consciousness.

We tend to approach wakefulness as an on-or-off phenomenon, a zero for sleep, a one for the awake state. To some extent that is correct, but the all-or-none approach hides gradations that we are all familiar with. Sleepiness and drowsiness certainly reduce consciousness, but they do not bring it to zero abruptly. Turning the lights off is not an accurate analogy; lowering a dimmer switch is closer to the mark.

What do the lights reveal when they are turned on, suddenly or gradually? More often than not, they reveal something that we commonly describe as a "mind" or "mental contents." And what is the mind so revealed made of? Patterns mapped in the idiom of every possible sense—visual, auditory, tactile, muscular, visceral, you name it, in marvelous shades, tones, variations, and combinations, flowing in orderly or jumbled manner, in brief, *images*. Earlier I offered my views on the origin of images (Chapter 3), and all we need to do here is recall that images are the main currency of our minds, and that the term refers to patterns of all sensory modalities, not just visual, and to abstract as well as concrete patterns.

Does the simple physiological act of turning on the lights—waking someone up from a slumber—necessarily translate into a conscious state? It definitely does not. We need not go very far to find counterevidence. Everyone has had the experience of waking up tired and jet-lagged, in some other land beyond the seas, and taking a thankfully short but seemingly long second or two to realize where exactly one is. During that brief interval there is a mind but not quite yet a mind organized with all the properties of consciousness. If I lose consciousness as a result of knocking my head against a less-than-soft object, I will have another blissfully short and yet measurable delay until "coming to." By the way, "coming to" is short for "coming to consciousness," returning to a self-oriented mind; the phrase is inelegant but celebrates a sound folk wisdom. In neurological jargon, regaining consciousness after a closed-head injury can take its sweet time, during which the victim is not fully oriented to place or clock, let alone to person.

Those situations show us that complex mental functions are not monoliths and can literally be broken down by sections. Yes, the lights are on and you are awake. (Strike one point for consciousness.) Yes, the mind is on, and images are being formed of whatever is in front of you, although those recalled from the past are few and far between. (Strike half a point for consciousness.) But no, there is little yet to indicate who the owner is of this shaky mind, no self to claim it as its own. (Strike no point for consciousness.) Overall, consciousness fails to pass. The moral of the story: to get a passing standard consciousness score, it is indispensable (1) to be awake; (2) to have an operational mind; *and* (3) to have, within that mind, an automatic, unprompted, undeduced sense of self as protagonist of the experience, no matter how subtle the self sense may be. Given the presence of wakefulness and mind, both of which you will need if you are to be conscious, you might say that the distinctive feature of your consciousness is, lyrically speaking, the very thought of you. But in order to make the poetry accurate, you would have to say "the very *felt* thought of you."

That wakefulness and consciousness are not one and the same is apparent when we consider the neurological condition known as vegetative state. Patients in a vegetative state have no manifestation suggestive of consciousness. Like patients in the similar but more grave situation of coma, vegetative patients fail to respond to any message from the examiners and offer no spontaneous signs of awareness of self or surroundings. And yet their electroencephalograms, or EEGs (the electrical wave patterns continuously produced by a living brain), reveal alternating patterns characteristic of either sleep or wakefulness. Along with wake-pattern EEGs, patients will often have their eyes open, although they stare vacantly into space, not directing their gaze to any particular object. No such electrical pattern is noted when patients are in coma, a situation in which all phenomena associated with consciousness (wakefulness, mind, and self) appear to be absent.[2]

The troubling condition of vegetative state also provides valuable information on another aspect of the distinctions I am drawing. In a study that justifiably attracted much attention, Adrian Owen was able to determine, using functional magnetic resonance imaging, that the brain of a woman in a vegetative state exhibited patterns of activity congruent with an examiner's questions and requests of her. Needless to say, she had been diagnosed as unconscious. She did not overtly respond to questions asked or directions offered, and she did not spontaneously give any evidence of active mind. And yet her fMRI study showed that the auditory regions of her cerebral cortices did become active when she was asked questions. The activation pattern resembled what one may see in a normal conscious subject responding to a comparable question. Even more impressive was the fact that when the patient was asked to imagine a tour of her own home, the cerebral cortices exhibited a pattern of activity of the sort one can find in normal conscious subjects doing a similar task. Although the patient did not reveal this exact same pattern on other occasions, a small number of other patients have since been studied in whom a comparable pattern was seen, though not in all attempts.[3] One of those patients, in particular, was able to evoke responses previously associated with *yes* or *no* by means of repeated training.[4]

The study indicates that even in the absence of all behavioral signs of consciousness, there can be signs of the kind of brain activity commonly correlated with mind processes. In other words, direct brain observations offer evidence compatible with some preservation of both wakefulness and mind, while behavioral observations reveal no evidence that consciousness, in the sense described earlier, accompanies such operations. These important results can be parsimoniously interpreted in the context of the abundant evidence that mind processes operate nonconsciously (as reviewed in this chapter and in Chapter 11). The findings are certainly compatible with the presence of a mind process and even a minimal self process. But in spite of the significance of these findings, scientifically and in terms of medical management, I am reluctant to regard them as evidence for conscious communication or as reasonable justification to abandon the definition of consciousness discussed earlier.

Removing the Self and Keeping a Mind

Perhaps the most convincing evidence for a dissociation between wakefulness and mind, on the one hand, and self, on the other, comes from another neurological condition, epileptic automatism, which can follow episodes of certain epileptic seizures. In such situations, a patient's behavior is suddenly interrupted for a brief period of time, during which the action freezes altogether; it is then followed by a period, generally brief as well, during which the patient returns to active behavior but gives no evidence of a normal conscious state. The silent patient may move about, but his actions, such as waving goodbye or leaving a room, reveal no overall purpose. The actions may exhibit a "minipurpose," like picking up a glass of water and drinking from it, but no sign that the purpose is part of a larger context. The patient makes no attempt to communicate with the observer and no reply to the observer's attempts.

If you visit a physician's office, your behavior is part of a large context that has to do with the specific goals of the visit, your overall plan for the day, and the wider plans and intentions of your life, at varied time scales, relative to which your visit may be of some significance or not. Everything you do in the "scene" at that office is informed by these multiple contents, even if you do not need to hold them all in mind in order to behave coherently. The same happens with the physician, relative to his role in the scene. In a state of diminished consciousness, however, all that background influence is reduced to little or nothing. The behavior is controlled by immediate cues, devoid of any insertion in the wider context. For example, picking up a glass and drinking from it makes sense if you are thirsty, and that action does not need to connect with the broader context.

I remember the very first patient I observed with this condition because the behavior was so new to me, so unexpected, and so disquieting. In the middle of our conversation, the patient stopped talking and in fact suspended moving altogether. His face lost expression, and his open eyes looked past me, at the wall behind. He remained motionless for several seconds. He did not fall from his chair, or fall asleep, or convulse, or twitch. When I spoke his name, there was no reply. When he began to move again, ever so little, he smacked his lips. His eyes shifted about and seemed to focus momentarily on a coffee cup on the table between us. It was empty, but still he picked it up and attempted to drink from it. I spoke to him again and again, but he did not reply. I asked him what was going on, and he did not reply. His face still had no expression, and he did not look at me. I called his name, and he did not reply. Finally he rose to his feet, turned around, and walked slowly to the door. I called him again. He stopped and looked at me, and a perplexed expression came to his face. I called him again, and he said, "What?"

The patient had suffered an absence seizure (a kind of epileptic seizure), followed by a period of automatism. He had been both there and not, awake and behaving, for sure, partly attentive, bodily present, but unaccounted for as a person. Many years later I described the patient as having been "absent without leave," and that description remains apt.[5]

Without question this man was awake in the full sense of the term. His eyes were open, and his proper muscular tone enabled him to move about. He could unquestionably produce actions, but the actions did not suggest an organized plan. He had no overall purpose and made no acknowledgment of the conditions of the situation, no appropriateness, and his acts were only minimally coherent. Without question his brain was forming mental images, although we cannot vouch for their abundance or coherence. In order to reach for a cup, pick it up, hold it to one's lips, and put it back on the table, the brain must form images, quite a lot of them, at the very least visual, kinesthetic, and tactile; otherwise the person cannot execute the movements correctly. But while this speaks for the presence of mind, it gives no evidence of self. The man did not appear to be cognizant of who he was, where he was, who I was, or why he was in front of me.

In fact, not only was evidence of such overt knowledge missing, but there was no indication of *covert* guidance of his behavior, the sort of nonconscious autopilot that allows us to walk home without consciously focusing on the route. Moreover, there was no sign of emotion in the man's behavior, a telltale indication of seriously impaired consciousness.

Such cases provide powerful evidence, perhaps the only definitive evidence yet, for a break between two functions that remain available, wakefulness and mind, and another function, self, which by any standard is not available. This man did not have a sense of his own existence and had a defective sense of his surroundings.

As so often happens when one analyzes complex human behavior that has been broken down by brain disease, the categories one uses to construct hypotheses regarding brain function and to make sense of one's observations are hardly rigid. Wakefulness and mind are not all-or-none "things." Self, of course, is not a thing; it is a dynamic process, held at some fairly stable levels during most of our waking hours but subject to variations, big and small, during that period, especially at the tail ends. Wakefulness and mind, as conceived here, are processes too, never rigid things. Turning processes into things is a mere artifact of our need to communicate complicated ideas to others, rapidly and effectively.

In the case just described, one can assume with confidence that wakefulness was intact and the mind process was present. But one cannot say how rich that mind process was, only that it was sufficient to navigate the limited universe the man was coping with. As for consciousness, it was clearly not normal.

How do I interpret the man's situation with the advantage of what I know today? I believe his assembling of a self function was severely compromised. He had lost the ability to generate, moment by moment, most of the self operations that would have given him, automatically, a proprietary survey of his mind. Those self operations would also have included elements of his identity, of his recent past and his intended future, and provided him with a sense of agency as well. The mental contents that a self process would have surveyed were probably impoverished. Under the circumstances, our man was confined to an aimless, unsituated now. The self as material me was mostly gone, and so was, even more certainly, the self as knower.

Being awake, having a mind, and having a self are different brain processes, concocted by the operation of different brain components. They merge seamlessly on any given day, in a remarkable functional continuum inside our brains, permitting and revealing different manifestations of behavior. But they are not "compartments" as such. They are not rooms divided by rigid walls because biological processes are not at all like artifacts engineered by humans. Still, in their messy, fuzzy, biological way, they are separable, and if we do not try to discover how they differ and where the subtle transitions occur, we have no prayer of understanding how the whole thing works.

I would say that if one is awake and there are contents in one's mind, consciousness is the result of adding a self function to mind that orients the mental contents to one's needs and thus produces subjectivity. The self function is not some know-all homunculus but rather an emergence, within the virtual screening process we call mind, of yet another virtual element: an imaged *protagonist* of our mental events.

Completing a Working Definition

When neurological disease breaks consciousness apart, emotional responses are notoriously absent, and the corresponding feelings are presumably missing as well. Patients with disturbances of consciousness fail to exhibit signs of ongoing emotion. Their faces have a blank, vacuous expression. Minor signs of muscular animation are absent, a remarkable feature given that even a so-called poker face is emotively animated and betrays subtle signs of expectation, glibness, contempt, and the like. Patients in any variant of akinetic mute or vegetative state, not to mention coma, have little or no emotional expression. The same is true of deep anesthesia but not, predictably, of sleep, in which emotional expressions may appear when the sleep stage permits paradoxical consciousness.

From a behavioral standpoint, the conscious mind state of others is hallmarked by awake, coherent, purposeful behavior that includes signs of ongoing emotional reactions. Very early in our lives we learn to confirm, based on the direct verbal reports we hear, that such emotional reactions are systematically accompanied by feelings. Later we assume, from looking at human beings around us, that they are experiencing certain feelings, even if they do not say a single word and no word is addressed to them. In fact, even the subtlest of emotional expressions can betray, to a well-tuned, syntonic, empathetic mind, the presence of feelings, no matter how quiet they may be. This process of feeling attribution has nothing whatsoever to do with language. It is based on the highly trained observation of postures and faces as they change and move about.

Why are emotions such a telltale sign of consciousness? Because the actual execution of most emotions is carried out by the periaqueductal gray (PAG) in close cooperation with the nucleus tractus solitarius (NTS) and the parabrachial nucleus (PBN), the structures whose ensemble engenders bodily feelings (such as primordial feelings) and the variations thereof that we call emotional feelings. This ensemble is often damaged by the neurological lesions that cause loss of consciousness, and certain anesthetics that target it can render it dysfunctional.

We shall see in the next chapter that just as signs of emotion are part of the externally observable conscious state, experiences of bodily feelings are a deep and vital part of consciousness from a first-person, introspective perspective.

Kinds of Consciousness

Consciousness fluctuates. Below a certain threshold consciousness is not operating, and along a scale of levels it operates in the most efficient way. Let us call this the "intensity" scale of consciousness, and let us exemplify those very different levels. In some moments you feel sleepy and are about to vanish into the arms of Morpheus; in another you are participating in an intense debate that calls for a keen awareness to the details that keep cropping up. The intensity scale ranges from dull to sharp, with all the shades in between.

Besides intensity, however, there is another criterion on which we can rate consciousness. It has to do with *scope*. Minimal scope allows a sensing of the self, say when one is drinking a cup of coffee at home, unconcerned with the provenance of either the cup or the coffee, or with what it will do to your heartbeat, or with what you have to do today. You are quietly present in the moment, that's all. Now suppose you are sitting down for a similar cup of coffee at a restaurant to meet with your brother, who wishes to discuss your parents' inheritance and what is to be done with your half sister, who has been acting strangely. You are still very present and in the moment, as they say in Hollywood, but now you are also transported, by turns, to many other places, with many other people besides your brother, and to situations that you have not experienced yet that are products of your informed and rich imagination. What your life has been, in bits and pieces, is available to you rapidly in recall, and bits and pieces of what your life may or not come to be, imagined earlier or imagined now, also come into the moment of experience. You are busily all over the place and at many epochs of your life, past and future. But you—the *me* in you, that is—never drops out of sight. All of these contents are inextricably tied to a singular reference. Even as you concentrate on some remote event, the connection remains. The center holds. This is big-scope consciousness, one of the grand achievements of the human brain and one of the defining traits of humanity. This is the kind of brain process that has brought us to where we are in civilization, for better and worse. This is the kind of consciousness illustrated by novels, films, and music and celebrated by philosophical reflection.

I have given names to those two kinds of consciousness. The minimal-scope kind I call *core* consciousness, the sense of the here and now, unencumbered by much past and by little or no future. It revolves around a core self and is about personhood but not necessarily identity. The big-scope kind I call *extended* or *autobiographical* consciousness, given that it manifests itself most powerfully when a substantial part of one's life comes into play and both the lived past and the anticipated future dominate the proceedings. It is about both personhood and identity. It is presided over by an autobiographical self.

More often than not, when we think about consciousness, we have in mind the broad-scope consciousness associated with an autobiographical self. Here the conscious mind widens and encompasses actual as well as imaginary contents effortlessly. Hypotheses regarding how the brain produces conscious states need to take into account this high level of consciousness as much as the core level.

Today I see the changes in consciousness scope as far more mercurial than I first envisioned them; that scope constantly shifts up or down a scale as if it moved on a gliding cursor. The upward or downward shift can occur *within* a given event, quite rapidly, as needed. This fluidity and dynamism regarding scope are not that different from the rapid shifting of intensity that is known to occur throughout the day and to which we already attended. When you are bored at a lecture, your consciousness is dulled and you may doze off and lose it. I sure hope it is not happening to you now.

By far the most important point to be made is that the levels of consciousness fluctuate with the situation. For instance, when I took my eyes off the page to think, and the dolphins that were swimming by caught my attention, I was not engaging the full scope of my autobiographical self because there would be no need for it; it would have been a waste of brain-processing capacity, not to mention fuel, given the needs of the moment. Nor did I need an autobiographical self to cope with the thoughts that preceded my writing of the preceding sentences. However, when an interviewer sits across from me and wants to know why and how I became a neurologist and neuroscientist rather than an engineer or filmmaker, I do need to engage my autobiographical self. My brain honors that need.

. . .

The level of consciousness also shifts rapidly when one daydreams, something that is now fashionably called mind-wandering. It might as well be called self-wandering because daydreaming requires not merely a lateral wandering away from the contents of the activity at hand but a downshift to core self. The products of our "offline" imagination move to the foreground—plans, occupations, fantasies, the sort of images that creep up when one is stuck on the Santa Monica Freeway. But online consciousness downshifted to core self and distracted to another topic is still normal consciousness. We cannot say the same about the consciousness of those who sleepwalk, or who are under hypnosis, or who experiment with "mind-altering" substances. Relative to the latter, the catalog of the resulting states of abnormal consciousness is long and varied and includes the most inventive aberrations of mind and self. Wakefulness breaks down as well, sleep or stupor being an all-too-common endpoint of such adventures.

In conclusion, the degree to which the protagonist self is present in our minds varies greatly with the circumstances, from a richly detailed and fully situated portrayal of who we are to an ever-so-faint hint that we do own our mind and our thoughts and our actions. But I must insist on the idea that even at its most subtle and faint, the self is a necessary presence in the mind. To say that when one is climbing a mountain, or when I am writing this very sentence, the self is nowhere to be found is not quite accurate. In such instances the self is not on prominent display for certain; it conveniently retreats to the background and makes room, in our image-making brain, for all the other things that require processing space—such as the face of the mountain or the thoughts I want to commit to the page. But I venture that if the self process were to collapse and disappear completely, the mind would lose its orientation, the ability to gather its parts. One's thoughts would be freewheeling, unclaimed by an owner. Our real-world efficacy would drop to little or nothing, and we would be lost for those observing us. What would we look like? Well, we would look unconscious.

I am afraid it is not easy dealing with the self because, depending on the perspective, the self can be so many things. It can be an "object" of research for psychologists and neuroscientists; it can be a provider of knowledge to the mind in which it emerges; it can be subtle and retreating behind a curtain or assertively present at the footlights; it can be confined to the here and now or encompass a whole life history; finally, some of these registers can be mixed, as when a knower self is subtle and yet autobiographic, or else prominently present but concerned only with the here and now. The self is indeed a movable feast.

Human and Nonhuman Consciousness

Just as consciousness is not a thing, the core and extended/auto-biographical kinds of consciousness are not rigid categories. I have always envisioned many grades between the core and autobiographical endpoints of the scale. But carving out these different kinds of consciousness has a practical payoff: it allows us to propose that the lower notches of the consciousness scale are by no means human alone. In all probability they are present in numerous nonhuman species that have brains complex enough to construct them. The fact that human consciousness, at its highest reaches, is hugely complicated, far-reaching, and therefore *distinctive* is so obvious that it does not require mention. The reader would be surprised, however, at how comparable comments of mine have, in the past, led some people to take offense, either because I was attributing too little consciousness to nonhuman species or because I was diminishing the exceptional nature of human consciousness by including animals. Wish me luck.

No one can prove satisfactorily that nonhuman, nonlanguaged beings have consciousness, core or otherwise, although it is reasonable to triangulate the substantial evidence we have available and conclude that it is highly likely that they do.

The triangulation would run like this: (1) if a species has behaviors that are best explained by a brain with mind processes rather than by a brain with mere dispositions for action (such as reflexes); and (2) if the species has a brain with all the components that are described in the chapters ahead as necessary to make conscious minds in humans; (3) then, dear reader, the species is conscious. At the end of the day, I am ready to take any manifestation of animal behavior that suggests the presence of feelings as a sign that consciousness should not be far behind.

Core consciousness does not require language and must have preceded language, obviously in nonhuman species but also in humans. In effect, language would likely not have evolved in individuals devoid of core consciousness. Why would they have needed it? On the contrary, at the highest grades on the scale, autobiographical consciousness relies extensively on language.

What Consciousness Is Not

Understanding the significance of consciousness, and the merits of its emergence in living beings, requires that we take a full measure of what came before, a sense of what living beings with normal brains and fully operational minds were capable of doing before their species came to have consciousness and before consciousness dominated mental life for those who had it. Watching the dissolution of consciousness in an epileptic patient or in someone in a vegetative state may give, to the unsuspecting observer, the erroneous notion that the processes that normally sit beneath consciousness are trivial or of limited effectiveness. But clearly the unconscious space of our own minds denies such an idea. I am referring here not just to the Freudian unconscious of famed (and infamous) tradition, identified with particular kinds of content, situation, and process. I am referring rather to the large unconscious that is made up of two ingredients: an active ingredient, constituted by all the images that are being formed on every topic and of every flavor, images that cannot possibly compete successfully for the favors of the self and therefore remain largely unknown; and a dormant ingredient, constituted by the repository of coded records from which explicit images can be formed.

A typical cocktail party phenomenon reveals the presence of the nonconscious quite well. While you are engaged in conversation with your host, you are technically *hearing* other conversations, a fragment here, a fragment there, at the edges of the stream of consciousness—the *main* stream, that is. But hearing does not mean listening, necessarily, let alone listening attentively and connecting with what is heard. And so you overhear many things that do not demand the services of your self. Then all of a sudden something clicks, some fragment of conversation joins others, and a sensible pattern emerges regarding some of those things that you were overhearing so loosely. At that instant you form a meaning that does "attract" the self and now literally takes you away from your host's last sentence. He notices your momentary distraction, by the way, and while fighting off the topic intruding into the river of your consciousness, you return to the gentleman's last point and lamely, apologetically, say, "I'm sorry; say again?"

As far as one can tell, the phenomenon is the consequence of several conditions. First, the brain constantly produces an overabundant quantity of images. What one sees, hears, and touches, along with what one constantly recalls—prompted by the new perceptual images as well as by no identifiable reason—is responsible for large numbers of explicit images, accompanied by an equally large retinue of other images relating to the state of one's body as all this image-making unfolds.

Second, the brain tends to organize this profusion of material much as a film editor would, by giving it some kind of coherent narrative structure in which certain actions are said to cause certain effects. This calls for *selecting* the right images and *ordering* them in a procession of time units and space frames. This is not an easy task, since not all images are equal, from the perspective of their owner. Some are more connected with one's needs than others and are thus accompanied by different feelings. Images are differently valued. Incidentally, when I say "the brain tends to organize," rather than "the self organizes," I do so on purpose. On some occasions the editing goes on naturally, with mini-mal self-imposed guidance. One's editing success, on such occasions, depends on how "well educated" our nonconscious processes have been by our own mature selves. I will return to this issue in the last chapter.

Third, only a small number of images can be displayed clearly at any given time because the image-making space is so scarce: only so many images can be active and thus potentially attended at any given moment. What this really means is that the metaphoric "screens" in which your brain displays the selected and time-ordered images are quite limited. In today's computer jargon, it means that the number of windows you can open on your screen is limited. (In the generation that has grown up multitasking, in the digital age, the upper limits of attention in the human brain are being rapidly raised, something that is likely to change certain aspects of consciousness in the not-too-distant future, if it has not done so already. Breaking the glass ceiling of attention has obvious advantages, and the associative abilities generated by multitasking are a terrific advantage; but there may be trade-off costs in terms of learning, memory consolidation, and emotion. We have no idea what these costs may be.)

These three constraints (abundance of images, tendency to organize them in coherent narratives, and scarcity of explicit display space) have prevailed for a long time in evolution and have required effective man-agement strategies to prevent them from damaging the organism in which they occur. Given that the making of images was naturally selected in evolution because images permit a more precise evaluation of the environment and a better response to it, the strategic management of images likely evolved bottom up, early on, well before consciousness did. The strategy was to select automatically those images that were most valuable for ongoing life management—precisely the same crite-rion presiding over the natural selection of the image-making devices. Especially valuable images, given their importance for survival, were "highlighted" by emotional factors. The brain probably achieves this highlighting by generating an emotional state that accompanies the image in a parallel track. The degree of emotion serves as a "marker" for the relative importance of the image. This is the mechanism described in the "somatic marker hypothesis."[6] The somatic marker does not need to be a fully formed emotion, overtly experienced as a feeling. (That is what a "gut feeling" is.) It can be a covert, emotion-related signal of which the subject is not aware, in which case we refer to it as a *bias*. The notion of somatic markers is applicable not just to high levels of cognition but to those earlier stages of evolution. The somatic marker hypothesis offers a mechanism for how brains would execute a value-based selection of images and how that selection would translate in edited continuities of images. In other words, the principle for the selection of images was connected to life-management needs. I suspect the same principle presided over the design of primordial narrative structures, which involved the organism's body, its status, its interactions, and its wanderings in the environment.

All of the above strategies, I submit, began to evolve long before there was consciousness, just as soon as enough images were being made, perhaps as soon as real minds first bloomed. The vast unconscious probably has been part of the business of organizing life for a long, long time, and the curious thing is that it is still with us, as the great subterraneam under our limited conscious existence.

Why did consciousness prevail, once it was offered to organisms as an option? Why were consciousness-making brain devices naturally selected? One possible answer, which we will consider at the end of the book, is that generating, orienting, and organizing images of the body and of the outside world in terms of the organism's needs, increased the likelihood of efficient life management and consequently improved the chances of survival. Eventually consciousness added the possibility of *knowing* about the organism's existence and about its struggles to stay alive. Of course, knowing depended not just on the creation and display of explicit images but on their storage in implicit records. Knowing connected the struggles of existence with a unified, identifiable organism. After such states of knowing began to be committed to memory, they could be connected to other recorded facts, and knowledge about individual existence could begin to be accumulated. In turn, the images contained in knowledge could be recalled and manipulated in a reasoning process that paved the way for reflection and deliberation. The image-processing machinery could then be guided by reflection and used for *effective anticipation of situations, previewing of possible outcomes, navigation of the possible future,* and *invention of management solutions.*

Consciousness allowed the organism to become cognizant of its own plight. The organism no longer had mere feelings that could be felt; it had feelings that could be *known,* in a particular context. Knowing, as opposed to being and doing, was a critical break.

Prior to the appearance of self and standard consciousness, organisms had been perfecting a machine of life regulation, on whose shoulders consciousness came to be built. Before some of the premises of the concern could be known in the conscious mind, those premises were already present, and the machine of life regulation had evolved around them. The difference between life regulation before consciousness and after consciousness simply has to do with automation versus deliberation. Before consciousness, life regulation was entirely automated; after consciousness begins, life regulation retains its automation but gradually comes under the influence of self-oriented deliberations.

Thus the foundations for the processes of consciousness are the unconscious processes in charge of life regulation—the blind dispositions that regulate metabolic functions and are housed in brain-stem nuclei and hypothalamus; the dispositions that deliver reward and punishment and promote drives, motivations, and emotions; and the mapping apparatus that manufactures images, in perception and recall, and that can select and edit such images in the movie known as mind. Consciousness is just a latecomer to life management, but it moves the whole game up a notch. Smartly, it keeps the old tricks in place and lets them do the journeymen jobs.

The Freudian Unconscious

Freud's most interesting contribution to consciousness comes from his very last paper, written in the second half of 1938 and left incomplete at the time of his death.[7] I read this paper only recently, prompted by an invitation to give a lecture on the topic of Freud and neuroscience. It is the sort of assignment one should decline vigorously, but I was tempted and accepted. I then spent weeks reviewing Freud's papers, alternating between irritation and admiration, as always happens when I read Freud. At the end of the toil came this final piece, which Freud wrote in London and in English, and where he adopts the only position on the matter of consciousness that I find plausible. Mind is a most natural result of evolution, and it is largely nonconscious, internal, and unrevealed. It comes to be known thanks to the narrow window of consciousness. This is precisely how I see it. Consciousness offers a direct experience of mind, but the broker of the experience is a self, which is an internal and imperfectly constructed informer rather than an external, reliable observer. The brain-ness of mind cannot be directly appreciated either by the natural internal observer or by the external scientist. The brain-ness of mind has to be imagined in the fourth perspective. Hypotheses have to be formulated on the basis of that imaginary view. Predictions have to be made on the basis of the hypotheses. A research program is needed to get closer to them.

Although Freud's view of the unconscious was dominated by sex, he was aware of the immense scope and power of mind processes going on under the sea level of consciousness. He was not alone, by the way, as the notion of unconscious processing was quite popular in psychological thinking of the last quarter of the nineteenth century. Nor was Freud alone in his foray into sex, whose science was also being explored at the time.[8]

Freud certainly seized on a wellspring of evidence for the unconscious when he concentrated on dreams. This move served his purposes quite well, as it provided him with material for his studies. This same wellspring has also been tapped by artists, composers, writers, and all manner of creators attempting to free themselves from the trammels of consciousness in search of novel images. A most interesting tension is at play here: very conscious creators consciously seek the unconscious as a source and, on occasion, as a method for their conscious endeavors. This in no way contradicts the idea that creativity could not have begun, let alone flourished, in the absence of consciousness. It just underscores how remarkably hybrid and flexible our mental lives are.

The reasoning of dreams is relaxed, to say the least, in the good dreams as well as in the nightmares, and while causality may be respected, the imagination goes wild and reality be damned. Dreams do offer, however, direct evidence of mind processes unassisted by consciousness. The depth of unconscious processing tapped by dreams is considerable. For those who may be reluctant to accept this, the most convincing instances may come from dreams that deal with plain life-regulation issues. A case in point: the person who dreams elaborately about fresh water and thirst after having had a dinner of very salty food. Ah, but wait!, I can hear the reader saying, what can you possibly mean when you say that the dream mind is "unassisted by consciousness"? Is it not the case that if one can remember a dream, then one was conscious when it happened? Well, that is indeed the case, in many instances. During dreams some kind of nonstandard consciousness is going on, the term *paradoxical* being quite apt. But my point is that the imaginative process depicted in dreams is not guided by a regular, properly functioning self of the kind we deploy when we reflect and deliberate. (The exception is the situation of so-called lucid dreaming, during which trained dreamers manage to self-direct their dreams to a certain extent.) Our mind, conscious as well as not, is probably *paced* by the outside world, whose inputs assist with the organization of contents. Deprived of that external pacemaker, it would be easy for the mind to dream itself away.[9]

The matter of remembering dreams is a vexing issue. We dream profusely, several times a night, when we are in rapid eye movement (REM) sleep, and we even dream, albeit far less so, when we are in slow-wave sleep, also known as non-REM (N-REM). But we seem to remember best the dreams that occur close to the return of consciousness as we ascend, gradually or not so gradually, to sea level.

I try hard to remember my dreams, but unless I write them down, they vanish without a trace, always did. It's not so surprising when we think that as we awaken, the memory-consolidation apparatus is barely on, like an oven in a bakery at first light.

The only type of dream I used to remember a bit better, perhaps because it was so well exercised, was a recurrent soft nightmare that would come the night before I was supposed to give a lecture. The variations always had the same gist: I am late, desperately late, and something essential is missing. My shoes may have disappeared; or my five o'clock shadow is turning into a two-day beard and my shaver is nowhere to be found; or the airport has closed down with fog and I am grounded. I am tortured and sometimes embarrassed, as when (in my dream, of course) I actually walked onstage barefoot (but in an Armani suit). That is why, to this day, I never leave shoes to be shined outside a hotel room.

8

Building a Conscious Mind

A Working Hypothesis

It goes without saying that the construction of a conscious mind is a very complex process, the result of additions and deletions of brain mechanisms over millions of years of biological evolution. No single device or mechanism can account for the complexity of the conscious mind. The different parts of the consciousness puzzle have to be treated separately and given their due before we can attempt a comprehensive account.

Still, it is helpful to start with a general hypothesis. The hypothesis comes in two parts. The first specifies that the brain constructs consciousness by generating a self process within an awake mind. The essence of the self is a focusing of the mind on the material organism that it inhabits. Wakefulness and mind are indispensable components of consciousness, but the self is the distinctive element.

The second part of the hypothesis proposes that the self is built in stages. The simplest stage emerges from the part of the brain that stands for the organism (the *protoself*) and consists of a gathering of images that describe relatively stable aspects of the body and generate spontaneous feelings of the living body (primordial feelings). The second stage results from establishing a relationship between the *organism* (as repre-

---

First stage: the protoself

the protoself is a neural description of relatively stable aspects of the organism

the main product of the protoself is spontaneous feelings of the living body (*primordial feelings*)

Second stage: the core self

a pulse of core self is generated when the protoself is modified by an interaction between the organism and an object and when, as a result, the images of the object are also modified

the modified images of object and organism are momentarily linked in a coherent pattern

the relation between organism and object is described in a narrative sequence of images, some of which are feelings

Third stage: the autobiographical self

the autobiographical self occurs when objects in one's biography generate pulses of core self that are, subsequently, momentarily linked in a large-scale coherent pattern

---

Figure 8.1: Three stages of self.

sented by the protoself) and any part of the brain that represents an *object-to-be-known*. The result is the *core self*. The third stage allows multiple objects, previously recorded as lived experience or as anticipated future, to interact with the protoself and produce an abundance of core self pulses. The result is the *autobiographical self*. All three stages are constructed in separate but coordinated brain workspaces. These are the image spaces, the playground for the influence of both ongoing perception and of dispositions contained in convergence-divergence regions.

. . .

By way of background and before presenting the several hypothetical mechanisms needed to carry out the general working hypothesis, let us say that, from an evolutionary standpoint, self processes began to occur only *after* minds and alertness were established as brain operations. Self processes were especially efficient at orienting and organizing minds toward the homeostatic needs of their organisms and thus increasing the chances of survival. Not surprisingly, self processes were naturally selected and prevailed in evolution. In early stages, the self processes probably did not generate consciousness in the full sense of the term and were confined to the protoself level. Later in evolution more complex levels of self—core self and beyond—began to generate subjectivity within the mind and to qualify for consciousness. Even later, ever more complex constructions were used to obtain and accumulate additional knowledge about individual organisms and about their environment. The knowledge was deposited in memories residing inside the brain, held in convergence-divergence regions and in memories that have been recorded externally, in the instruments of culture. Consciousness in the fullest sense of the term emerged after such knowledge was categorized, symbolized in varied forms (including recursive language), and manipulated by imagination and reason.

Two additional qualifications are in order. First, distinct levels of processing—mind, conscious mind, and conscious mind capable of producing culture—emerged in sequence. That should not leave the impression, however, that when minds acquired selves, they stopped evolving as minds or that selves eventually stopped evolving. On the contrary, the evolutionary process continued (and continues), possibly enriched and accelerated by the pressures created by self-knowledge, and there is no end in sight. The ongoing digital revolution, the globalization of cultural information, and the coming of the age of empathy are pressures likely to lead to structural modifications of mind and self, by which I mean modifications of the very brain processes that shape the mind and self.

Second, from this point forward in the book, we will approach the problem of building a conscious mind from the perspective of the human, although whenever possible and appropriate, reference will be made to other species.

Approaching the Conscious Brain

The neuroscience of consciousness is often approached from the mind component rather than from the self.[1] Opting to approach consciousness via the self is not meant to diminish, let alone neglect, the complexity and scope of sheer minds. Giving pride of place to the self process, however, is in keeping with the perspective adopted at the outset, according to which the reason why conscious minds prevailed in evolution was the fact that consciousness optimized life regulation. The self in each conscious mind is the first representative of individual life-regulation mechanisms, the guardian and curator of biological value. To a considerable extent, the immense cognitive complexity that hallmarks the current conscious minds of humans is motivated and orchestrated by the self, as a proxy of value.

Whatever one's study preference may be regarding the triad of wakefulness, mind, and self, it is apparent that the mystery of consciousness does not reside with wakefulness. On the contrary, we have considerable knowledge about the neuroanatomy and neurophysiology behind the process of wakefulness. It is perhaps no coincidence that the history of research on brain and consciousness actually began with the matter of wakefulness.[2]

The mind is the second component in the consciousness triad, and regarding its neural basis we are not in the dark either. We have made some progress, as discussed in Chapter 3, even if many questions remain. That leaves the third and central component of the triad, the self, whose approach is often postponed on the grounds that it is too complicated to tackle in the current stage of our knowledge. This chapter and the next are largely concerned with the self, and they outline mechanisms for generating it and inserting it into the awake mind. The goal is to identify the neural structures and the mechanisms that are possibly capable of producing self processes, ranging from the sort of simple self that orients behavior adaptively to the complex variety of self capable of knowing that its own organism exists and guiding life accordingly.

Previewing the Conscious Mind

Among the self's many levels, the most complex tend to obscure the view of the simpler ones, dominating our minds with an exuberant display of knowledge. But we can try to overcome the natural obfuscation and put all this complexity to a good use. How? By asking the complex levels of self to *observe* what is going on at the simpler ones. This is a difficult exercise and not without risks. Introspection, as we have seen, can provide misleading information. But the risk is well worth taking, given that introspection offers the only direct view of what we wish to explain. Besides, if the information we gather leads to flawed hypotheses, then future empirical testing will reveal them to be so. On an intriguing note, engaging in introspection turns out to be a translation, within the mind, of a process that complex brains have been engaged in for a long time in evolution: talking to themselves, both literally and in the language of neuron activity.

Let us look, then, inside our conscious minds and try to observe what the mind is like, at the bottom of its richly layered textures, stripped of the baggage of identity, lived past, and anticipated future, the conscious mind of the moment and in the moment. I cannot speak for everyone, of course, but here is what my reconnaissance tells me. To begin with, down at the bottom, the simple conscious mind is not unlike what William James described as a flowing stream with objects in it. But the objects in the stream are not equally salient. Some are as if magnified, others not. Nor are the objects arranged equally relative to me. Some are placed in a certain perspective relative to a material me that, a good part of the time, I can even localize not just to my body but, more precisely, to a bit of space behind my eyes and between my ears. Just as notably, some objects, though not all, are accompanied by a feeling that connects them unequivocally with my body and mind. The feeling tells me, without a word being spoken, that I own the objects, for the duration, and that I can act on them if I wish to do so. This is, literally, "the feeling of what happens," the object-related feeling about which I have written in the past. On the matter of feelings in the mind, however, I have this to add: *the feeling of what happens is not the whole story.* There is some deeper feeling to be guessed and then found in the depths of the conscious mind. It is the feeling that my own body exists, and it is present, independently of any object with which it interacts, as a rock-solid, wordless affirmation that I am alive. This fundamental feeling, which I had not deemed necessary to note in earlier approaches to this problem, I now introduce as a critical element of the self process. I call it *primordial feeling,* and I note that it has a definite *quality,* a *valence,* somewhere along the pleasure-to-pain range. It is the primitive behind all feelings of emotion and therefore is the basis of all feelings caused by interactions between objects and organism. As we shall see, primordial feelings are produced by the protoself.[3]

In brief, while plunging into the depths of the conscious mind, I discover that it is a composite of different images. One set of those images describes the *objects* in consciousness. Other images describe *me,* and the *me* includes: (1) the *perspective* in which the objects are being mapped (the fact that my mind has a standpoint of viewing, touching, hearing, and so on, and that the standpoint is my body); (2) the feeling that the objects are being represented in a mind belonging to me and to no one else (*ownership*); (3) the feeling that I have *agency* relative to the objects and that the actions being carried out by my body are commanded by my mind; and (4) *primordial feelings,* which signify the existence of my living body independently of how objects engage it or not.

The aggregate of elements (1) through (4) constitutes a self in its simple version. When the images of the self aggregate are folded together with the images of nonself objects, the result is a conscious mind.

All this knowledge is readily present. It is not arrived at by reasoned inference or interpretation. To begin with, it is not verbal either. It is made of hints and hunches, of feelings that occur *relative to the living body* and *relative to an object*.

The simple self at the bottom of the mind is a lot like music but not yet poetry.

The Ingredients of a Conscious Mind

The basic ingredients in the construction of conscious minds are *wakefulness* and *images.* On the matter of *wakefulness,* we know that it depends on the operation of certain nuclei in the brain-stem tegmentum and the hypothalamus. Using both neural and chemical routes, these nuclei exert their influence on the cerebral cortex. As a result, vigilance is either diminished (producing sleep) or enhanced (producing wakefulness). The work of the brain-stem nuclei is assisted by the thalamus, although some nuclei influence the cerebral cortex directly; as for the hypothalamic nuclei, they operate largely by release of chemical molecules that subsequently act on neural circuits and alter their behavior.

The delicate balance of wakefulness depends on the close interplay of hypothalamus, brain stem, and cerebral cortex. The function of the hypothalamus is closely related to the amount of light available, the part of the wakefulness process whose disruption causes jet lag when we fly across several time zones. In turn, this operation is closely coupled with hormonal secretion patterns tied in part to day-night cycles. The hypothalamic nuclei control the operation of endocrine glands throughout the organism—pituitary, thyroid, adrenals, pancreas, testes, ovaries.[4]

The brain-stem component of the wakefulness process relates to the natural value of each ongoing situation. Spontaneously and nonconsciously, the brain stem answers questions that no one poses, such as, how much should the situation matter to the beholder? Value determines the signal and degree of emotional responses to a situation as well as how awake and alert we are to be. Boredom plays havoc with wakefulness, but so do metabolic levels. We know what happens during digestion of a large meal, especially if certain chemical ingredients are present such as tryptophan, which is released from red meats. Alcohol increases wakefulness at first, only to induce sleepiness later, as blood alcohol levels rise. Anesthetics suspend wakefulness altogether.

One last cautionary note regarding wakefulness: the sector of the brain stem involved in wakefulness is distinct, in terms of its neuroanatomy and neurophysiology, from the sector of the brain stem that generates the foundations of the self, the protoself (which is discussed in the next section). The *brain-stem wakefulness nuclei* are anatomically close to the *brain-stem protoself nuclei* for a very good reason: both sets of nuclei participate in life regulation. However, they contribute to the regulatory process in different ways.[5]

On the matter of *images,* it may seem that we already know what we need to know, given that we discussed their neural basis in Chapters 3 through 6. But we need to say more. Images are certainly the source of the *objects-to-be-known* in the conscious mind, whether the objects are out in the world (external to the body) or inside the body (like my painful elbow or the finger you burned inadvertently). Images come in all sensory varieties, not just visual, and they pertain to *any object or action being processed in the brain,* actually present or being recalled, concrete as well as abstract. This covers all patterns originating *outside the brain,* either inside or external to the body. It also covers patterns generated inside the brain as a result of conjunctions of other patterns. Indeed, the brain's ravenous map-making addiction leads it to map its own workings—once again, talk to itself. The brain's maps of its own doings are probably the main source of abstract images that describe, for example, spatial placements and movement of objects, relationships of objects, velocity and spatial course of objects in motion, and patterns of occurrence of objects in time and space. These sorts of images can be converted into mathematical descriptions as well as musical compositions and executions. Mathematicians and composers excel at this sort of image-making.

The working hypothesis advanced earlier proposes that conscious minds arise from establishing a *relationship* between the *organism* and an *object-to-be-known*. But how are the organism and the object and the relationship implemented in the brain? All three components are made of images. The object to be known is mapped as an image. So is the organism, although its images are special. As for the knowledge that constitutes a self state and permits the emergence of subjectivity, it too is made of images. The entire fabric of a conscious mind is created from the same cloth—*images generated by the brain's map-making abilities.*

Even though all aspects of consciousness are constructed with images, not all images are born equal in terms of neural origin or physiological characteristics (see Figure 3.1). The images used to describe most objects-to-be-known are conventional, in the sense that they result from the mapping operations we discussed for the external senses. But the images that stand for the organism constitute a particular class. They originate in the body's interior and represent aspects of the body in action. They have a special status and a special achievement: they are *felt,* spontaneously and naturally, from the get-go, prior to any other operation involved in the building of consciousness. They are *felt* images of the body, primordial bodily feelings, the primitives of all other feelings, including feelings of emotions. Later we shall see that the images that describe the relationship between organism and object draw on both kinds of images—conventional sensory images and variations on bodily feelings.

Finally, all images occur in an aggregate workspace that is formed by separate early sensory regions of the cerebral cortices and, in the case of feelings, by selected regions of the brain stem. This image space is controlled by a number of cortical and subcortical sites whose circuits contain dispositional knowledge recorded in dormant form in the convergence-divergence neural architecture we discussed in Chapter 6. The regions can operate either consciously or nonconsciously, but in either case they do so within precisely the same neural substrates. The difference between the conscious and unconscious modes of operation in the participating regions depends on degrees of wakefulness and on the level of self processing.

In terms of its neural implementation, the notion of image space advanced here differs considerably from the notions found in the work of Bernard Baars, Stanislas Dehaene, and Jean-Pierre Changeux. Baars originated the notion of global workspace, in purely psychological terms, to call attention to the intense cross-communication of different components of the mind process. Dehaene and Changeux came to use global workspace, in neuronal terms, to refer to the highly distributed and interrelational neural activity that must underlie consciousness. Brainwise, they focus on the cerebral cortex as a provider of contents of consciousness, and they privilege the association cortices, especially the prefrontal, as a necessary element in the access to those contents. Later work by Baars also puts the global workspace notion at the service of *access* to contents of consciousness.

For my part, I focus on the image-making regions, the playground where the puppets in the show actually play. The puppeteers and the strings are outside the image space, in dispositional space located in the association cortices of the frontal, temporal, and parietal sectors. This perspective is compatible with imaging studies and electrophysiological studies that describe the behavior of those two distinct sectors (image space and dispositional space) in relation to conscious versus nonconscious images, such as in the work of Nikos Logothetis or Giulio Tononi on binocular rivalry, or the work of Stanislas Dehaene and Lionel Naccache on word processing. Conscious states require early sensory engagement *and* the engagement of association cortices, because, as I see it, that is from where the puppet masters organize the show.[6] I believe my account of the problem complements the global neuronal workspace approach, rather than standing in conflict with it.

The Protoself

The protoself is the stepping-stone required for the construction of the core self. It is *an integrated collection of separate neural patterns that map, moment by moment, the most stable aspects of the organism's physical structure.* The protoself maps are distinctive in that they generate not merely body images but also *felt* body images. These primordial feelings of the body are spontaneously present in the normal awake brain.

The contributors to the protoself include *master interoceptive maps, master organism maps,* and *maps of the externally directed sensory portals.* From an anatomical standpoint, these maps arise both from the brain stem and from the cortical regions. The basic state of the protoself is an average of its interoceptive component and its sensory portals component. The integration of all these diverse and spatially distributed maps takes place by cross-signaling within the same time window. It does not require a single brain site where the diverse components would be remapped. Let us consider each of the protoself contributors individually.

MASTER INTEROCEPTIVE MAPS

These are the maps and images whose contents are assembled from the interoceptive signals that hail from the internal milieu and the viscera. The interoceptive signals tell the central nervous system about the ongoing state of the organism, which may range from the optimal or the routine to the problematic, when the integrity of an organ or tissue has been violated and damage has occurred in the body. (I am referring here to nociceptive signals, which are the basis of feelings of pain.) Interoceptive signals signify the need for physiological corrections, something that materializes in our minds, for example, like feelings of

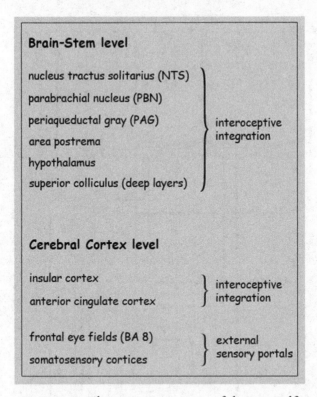

Figure 8.2: The main components of the protoself.

hunger and thirst. All the signals that convey temperature, along with myriad parameters of the operation of the internal milieu, are covered under this heading. Last, interoceptive signals participate in the making of hedonic states and the corresponding feelings of pleasure.

On any given moment, a subset of these signals, as assembled and modified in certain upper-brain-stem nuclei, generate primordial feelings. The brain stem is not a mere pass-through of the body signals to the cerebral cortex. It is a decision station, capable of sensing changes and responding in predetermined but modulated ways, at that very level. The workings of that decision machinery contribute to the *construction* of primordial feelings, so that such feelings are more than simple "portrayals" of the body, more elaborate than straightforward maps. Primordial feelings are a by-product of the particular way in which the

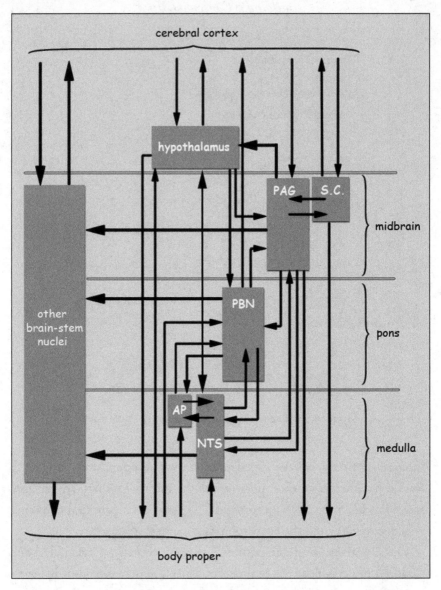

Figure 8.3: The brain-stem nuclei involved in generating the core self. As shown in Figure 4.1, several brain-stem nuclei work together to ensure homeostasis. But the homeostasis-related nuclei project to *other* groups of brain-stem nuclei (*other brain-stem nuclei,* in this figure). These other nuclei are grouped in functional families: the *classical nuclei of the reticular formation,* such as the nucleus pontis oralis and nucleus cuneiform, which influence the cerebral cortex via the intralaminar nuclei of the thalamus; the *monoaminergic nuclei,* which directly release

molecules such as noradrenalin, serotonin, and dopamine to wide-spread regions of the cerebral cortex; and the *cholinergic nuclei,* which release acetylcholine.

In the hypothesis advanced here, the homeostatic nuclei generate the "feelings of knowing" component of the core self. In turn, the neural activity underlying that process recruits the other, nonhomeostatic brain-stem nuclei, to generate "object saliency."

Abbreviations are as in Figure 4.1.

brain-stem nuclei are organized and of their unbreakable loop with the body. The functional characteristics of the particular neurons involved in the operation possibly contribute as well.

Primordial feelings precede all other feelings. They refer specifically and uniquely to the living body that is interconnected with its specific brain stem. All feelings of emotion are variations of the ongoing primordial feelings. All feelings caused by the interaction of objects with the organism are variations of the ongoing primordial feelings. Primordial feelings and their emotional variations generate an observant chorus that accompanies all other images going on in the mind.

The importance of the interoceptive system for the understanding of the conscious mind cannot be emphasized enough. The processes in this system are largely *independent* of the size of the structures in which they arise, and they constitute a special kind of input that is present from early on in development and throughout childhood and adolescence. In other words, interoception is a suitable source for the relative *invariance* required to establish some sort of stable scaffolding for what will eventually constitute the self.

The issue of relative invariance is critical because the self is a singular process and we must identify a plausible biological means to ground that singularity. On the face of it, the organism's single body should provide that much-needed biological singularity. We live in one body, not in two (not even Siamese twins deny this fact), and we have one mind to go with that body and one self to go with both. (Multiple selves and multiple personalities are not normal states of mind.) But the single grounding platform cannot possibly correspond to the *whole body* because, as a whole, the body is continuously performing different actions and changing shape accordingly, not to mention growing in size from birth to adulthood. The single platform must be found elsewhere, in a part of the body *within* the body, rather than in the body as a unit. It must correspond to the sectors of the body that change the least or not at all. The internal milieu and many visceral parameters associated with it provide the most invariant aspects of the organism, at any age, across a lifetime, not because they do not change but because their operations require that their condition vary only within an extremely narrow range. Bones grow across development time, and so do the muscles that move them; but the essence of the chemical bath in which life occurs— the average range of its parameters—is approximately the same whether you are three years old or fifty or eighty. Also, whether one is two feet tall or six, the biological essence of a state of fear or happiness is in all likelihood the same in terms of how such states are constructed from the chemistries in the internal milieu and the state of contraction or dilation of smooth muscles in the viscera. It is worth noting that the causes of a state of fear or happiness—the thoughts that cause those states—may be quite different across a lifetime, but the profile of one's emotional reaction to those causes is not.

Where does the master interoceptive system operate? The answers have become quite elaborate over the past decade thanks to work ranging from physiological recordings at the cellular level and experimental neuroanatomy studies in animals, to functional neuroimaging in humans. The upshot of this research (discussed in Chapter 4) is some unusually detailed knowledge about the pathways that bring such signals to the central nervous system.[7] The neural and chemical signals that describe body states enter the central nervous system at many levels of the spinal cord, the trigeminal nucleus in the brain stem, and the special collections of neurons that hover on the margin of the brain's ventricles. From all the entry points, the signals are relayed to major integrative nuclei in the brain stem; the most important are in the nucleus tractus solitarius, the parabrachial nucleus, and the hypothalamus. From there, after being processed locally and used to regulate the life process and generate primordial feelings, they are *also* relayed to the sector most clearly identified with interoception, the insular cortex, after a convenient stop in the thalamic relay nuclei. Notwithstanding the significance of the cortical component of this system, I see the brain-stem component as foundational for the self process. It can provide an operational protoself as specified in the hypothesis, even when the cortical component is extensively compromised.

MASTER ORGANISM MAPS

The master organism maps describe a schema of the entire body with its major components—head, trunk, and limb—in repose. The movements of the body are mapped against that master map. Unlike the interoceptive maps, the master organism maps change dramatically during development because they portray the musculoskeletal system and its motion. Of necessity, these maps follow increases in body size and in the range and quality of motion. They could not conceivably be the same in a toddler, an adolescent, and an adult, although some sort of temporary stability is eventually reached. As a result, the master organism maps are not the ideal source of the singularity required to constitute the protoself.

The master interoceptive system must fit within the general framework created by the master organism schema, at every phase of the latter's growth. A rough sketch would depict the master interoceptive system *within* the perimeter of the master organism framework. But the two are distinct. The fit of one system into the other does not imply an actual transfer of maps but rather a coordination such that both sets of maps can be evoked at the same time. For example, the mapping of a specific region of the body's interior would be signaled to the sector of the master organism framework where the region best fits into the overall anatomical scheme. When we sense nausea, we often experience it in relation to a region of the body—the stomach, for example. In spite of its vagueness, this interoceptive map is made to fit into the overall organism map.

MAPS OF THE EXTERNALLY DIRECTED SENSORY PORTAL

I referred indirectly to the sensory portals in Chapter 4, by describing the armature into which the sensory probes—the diamonds—are set. Here I place them at the service of the self. The representation of the varied sensory portals in the body—like the body regions encasing the eyes, ears, tongue, nose—is a separate and special case of a master organism map. I imagine that sensory portal maps "fit" into the framework of the master organism maps much as the master feeling system must, by means of time coordination rather than by actual map transfer. Where exactly some of these maps are is a matter of current investigation.

Sensory portal maps play a dual role, first in the building of perspective (a major aspect of consciousness) and then in the construction of qualitative aspects of mind. One of the curious aspects of our awareness of an object is the exquisite relation we establish between the mental contents that describe the object and those that correspond to the body part engaged in the respective perception. We know that we see with our eyes, but *we also feel ourselves seeing with our eyes*. We know that we hear with our ears, not with our eyes or nose. We do feel sound in the external ear and tympanic membrane. We touch with our fingers and smell with our noses and so forth. This may sound trivial at first glance, but it is anything but. We know all of this "sense organ location" from a tender age, probably before we discover it by inference, connecting a certain perception with a particular movement, perhaps even before countless rhymes and songs instruct us, at school, on where the senses get their information. Nonetheless, this is an odd sort of knowledge.

Consider that visual images come from the neurons in the retina, which are not presumed to tell us anything about the sector of the body where the retinas happen to be located—inside the eyeballs, which are inside the eye sockets, within a specific part of the face. How did we ever find out that the retinas are where they are? Of course, a child will have noted that vision goes away when the eyes are closed and that closing the ears reduces hearing. But that is hardly the point. The point is that we "feel" sound coming into the ears, and we "feel" that we are looking around and seeing with our eyes. A child in front of a mirror would confirm knowledge that would already have been acquired thanks to adjunct information originating from body structures "around" the retina. The ensemble of those body structures constitutes what I call a *sensory portal*. In the case of vision, the sensory portal includes not only the eye musculature with which we move the eyes but also the entire apparatus with which we focus on an object by adjusting the size of the lens; the apparatus of light-intensity adjustment that reduces or increases the diameter of the pupils (the camera shutters of our eyes); and, finally, the muscles *around* the eyes, those with which we can frown, or blink, or signify mirth. Eye movements and blinking play a critical role in the editing of our own visual images, and remarkably they also play a role in the effective and realistic editing of film images.

Seeing consists of more than getting the appropriate light pattern on the retina. Seeing encompasses all these other co-responses, some of which are indispensable to generating a clear pattern in the retina, some of which are habitual accompaniments of the process of seeing, and some of which are already fast reactions to processing the pattern itself.

The case for hearing is comparable. The vibration of the tympanic membrane and of a set of minuscule bones in the middle ear can be signaled to the brain in parallel with the sound itself, which occurs in the internal ear, at the level of the cochleas, where sound frequencies, time, and timbre are mapped.

The complex operation of the sensory portals may contribute to the errors that children as well as adults can commit regarding the perception of an event—for example, reporting that a certain object was first seen and then heard, when the opposite happened. The phenomenon is known as source misattribution error.

The unsung sensory portals play a crucial role in defining *the perspective* of the mind relative to the rest of the world. I am not talking here about the biological singularity provided by the protoself. I am referring to an effect we all experience in our minds: having a *standpoint* for whatever is happening outside the mind. This is not a mere "point of view," although for the sighted majority of human beings, the view does dominate the proceedings of our mind, more often than not. But we also have a standpoint relative to the sounds out in the world, a standpoint relative to the objects we touch, and even a standpoint for the objects we feel in our own body—again, the elbow and its pain, or our feet as we walk on the sand.

We do not mistakenly think we see with our belly buttons or hear with our armpits (intriguing as these possibilities might be). The sensory portals near which the data for making images are collected provide the mind with the standpoint of the organism relative to an object. The standpoint is drawn from the collection of body regions around which perceptions arise. That standpoint is broken only in abnormal conditions (out-of-body experiences), which can result from brain disease, psychological trauma, or experimental manipulations using virtual reality devices.[8]

I envision organism perspective as grounded in a variety of sources. Sight, sound, spatial balance, taste, and smell all depend on sensory portals not far from one another, all located in the head. We can think of the head as a multidimensional surveillance device, ready to take in the world. Touch, in its all-overness, has a broader sensory portal, but perspective related to touch still points unequivocally to the singular organism as the surveyor, and it identifies a place on the surveyor's surface. The same all-overness obtains for the perception of our own movement, which does relate to the entire body but always originates with the singular organism.

As far as the cerebral cortex is concerned, most of the sensory portal data must land in the somatosensory system—with SI and SII favored over the insula. In the case of vision, sensory portal data are also conveyed to the so-called frontal eye fields, which are located in Brodmann's area 8, in the superior and lateral aspects of the frontal cortex. Once again these geographically separate brain regions need to be brought together functionally by some sort of integrating mechanism.

One last note is in order regarding the exceptional situation of somatosensory cortices. These cortices convey signals from the external world, touch maps being the prime example, and from the body, as in the case of interoception, and the sensory portals. The sensory portal component rightfully belongs to organism structure and thus to the protoself.

There is a remarkable contrast between two distinct sets of patterns, then. On the one hand, there is the infinite variety of patterns describing conventional objects (some of which are external to the body, such as sights and sounds, tastes and odors; some of which are actual body parts, such as joints or patches of skin). On the other hand, there is the infinite sameness of the narrow range of patterns related to the body's interior and its tightly controlled regulation. There is an inescapable and fundamental difference between the strictly controlled aspect of the life process present inside our organisms and all the imaginable things and events out in the world or in the rest of the body. This difference is indispensable to understanding the biological foundation of the self processes.

This same contrast between variety and sameness also holds at the level of the sensory portals. The changes that the sensory portals undergo from their basal state to the state associated with looking and seeing do not have to be extensive, although they can be. The changes simply have to signify that an engagement of organism and object has taken place. They do not have to convey anything about the object being engaged.

In brief, the combination of the internal milieu, the visceral structure, and the basal state of the externally directed sensory portals provides an island of stability within a sea of motion. It preserves a relative
coherence of functional state within a surround of dynamic processes
whose variations are quite pronounced. Picture a large crowd marching
along a street; a small group in the middle of the crowd is moving in
steady and cohesive formation, while the rest of the crowd is darting
loosely, in Brownian motion, some elements dragging behind the others, some overtaking the core group, and so forth.

One other element must to be added to the scaffolding provided by
the internal milieu's relative invariance: the fact that the body proper
remains inseparably attached to the brain at all times. This attachment
underlies the generation of primordial feelings and the unique relationship between the body, as object, and the brain that represents that
object. When we make maps of objects and events out in the world,
those objects and events remain out in the world. When we map our
body's objects and events, they are inside the organism and they do not
go anywhere. They act on the brain but can be acted upon at any time,
forming a resonating loop that achieves something akin to a body-mind
fusion. They constitute an animated substrate that provides an obligate
context for all other contents of the mind. The protoself is not a mere
collection of maps of the body comparable to the nice collection of pictures of abstract expressionist paintings that I carry in my brain. The
protoself is a collection of maps that remains connected interactively
with its source, a deep root that cannot be alienated. Alas, the pictures
of favorite abstract expressionist paintings that I carry in my brain do
not connect physically at all with their sources. I wish they did, but they
are only in my brain.

Finally, I should note that the protoself is not to be confused with a
homunculus, just as the self that results from its modification is not
homuncular. The traditional notion of homunculus corresponds to a

little person sitting inside the brain, all knowing and all wise, capable of answering questions about what is going on in the mind and providing interpretations for the goings-on. The well-identified problem with the homunculus resides with the infinite regress it creates. The little person whose knowledge would render us conscious needs to have yet another little person in its inside, capable of providing it with the necessary knowledge, and so forth ad infinitum. This does not work. The knowledge that renders our minds conscious must be constructed from the bottom up. Nothing could be farther from the notion of protoself presented here than the idea of homunculus. The protoself is a reasonably stable platform and thus a source of continuity. We use the platform to inscribe the changes caused by having an organism interact with its surround (as when one looks at and grasps an object) or to inscribe the modification of organism structure or state (as when one suffers a wound or lowers the level of blood sugar excessively). The changes are *registered against the current state of the protoself,* and the perturbation triggers subsequent physiological events, but the protoself does not contain any information aside from that contained in its maps. The protoself is not a sage sitting at Delphi answering questions about who we are.

Constructing the Core Self

In thinking about a strategy to construct the self, it is appropriate to start with the requirements for the core self. The brain needs to introduce into the mind something that was not present before, namely, a protagonist. Once a protagonist is available in the midst of other mind contents, and once that protagonist is coherently linked to some of the current mind contents, subjectivity begins to inhere in the process. We need to concentrate first on the protagonist's threshold, the point at which the indispensable elements of knowledge agglutinate, so to speak, to yield subjectivity.

. . .

Once we have a unified island of relative stability corresponding to a part of the organism, might the self emerge from it in one fell swoop? If so, the anatomy and physiology of the brain regions that underlie the protoself would tell most of the story of how a self is made. The self would derive from the brain's capacity to accumulate and integrate knowledge about the most stable aspects of the organism, case closed. The self would amount to the unadorned and *felt* representation of life within the brain, a sheer experience unconnected to anything but its own body. The self would consist of the primordial feeling that the protoself, in its native state, spontaneously and relentlessly delivers, instant after instant.

When it comes to the complex mental lives that both you and I are experiencing at this very moment, however, protoself and primordial feeling are not enough to account for the self phenomenon we are generating. The protoself and its primordial feelings are the likely foundation of the material me and are, in all probability, an important and peak manifestation of consciousness in numerous living species. But we need some intermediate self process placed between the protoself and its primordial feelings, on the one hand, and the autobiographical selves that give us our sense of personhood and identity, on the other. Something critical must change in the very state of the protoself for it to become a self in the proper sense, that is, a *core self*. For one thing, the mental profile of the protoself must be raised and made to *stand out*. For another, it must *connect* with the events that it is involved in. Within the narrative of the moment, it must *protagonize*. As I see it, the critical change of the protoself comes from its moment-to-moment engagement as caused by any object being perceived. The engagement occurs in close temporal proximity to the sensory processing of the object. Anytime the organism encounters an object, any object, the protoself is changed by the encounter. This is because, in order to map the object, the brain must adjust the body in a suitable way, and because the results of those adjustments as well as the content of the mapped image are signaled to the protoself.

Changes in the protoself inaugurate the momentary creation of the core self and initiate a chain of events. The first event in the chain is a transformation in the primordial feeling that results in a "feeling of knowing the object," a feeling that differentiates the object from other objects of the moment. The second event in the chain is a consequence of the feeling of knowing. It is a generation of "saliency" for the engaging object, a process generally subsumed by the term *attention,* a drawing in of processing resources toward one particular object more than others. The core self, then, is created by linking the modified protoself to the object that caused the modification, an object that has now been hallmarked by feeling and enhanced by attention.

At the end of this cycle, the mind includes images regarding a simple and very common sequence of events: an object engaged the body when that object was looked at, touched, or heard, from a specific perspective; the engagement caused the body to change; the presence of the object was felt; the object was made salient.

The nonverbal narrative of such perpetually occurring events spontaneously portrays in the mind the fact that there is a protagonist to whom certain events are happening, that protagonist being the material me. The portrayal in the nonverbal narrative simultaneously creates and reveals the protagonist, connects the actions being produced by the organism to that same protagonist, and, along with the feeling generated by engaging with the object, engenders a sense of ownership.

What is being added to the plain mind process and is thus producing a conscious mind is a series of images, namely, an *image* of the organism (provided by the modified protoself proxy); the *image* of an object-related emotional response (that is, a feeling); and an *image* of the momentarily enhanced causative object. *The self comes to mind in the form of images, relentlessly telling a story of such engagements.* The images of the modified protoself and of the feeling of knowing do not even have to be especially intense. They just have to be there in the mind, however subtly, little more than hints, to provide a connection between object and organism. After all, it is the object that most matters in order for the process to be adaptive.

I see this wordless narrative as an account of what is transpiring, in life as well as in the brain, but not yet as an interpretation. It is, rather, an unsolicited description of events, the brain indulging in answering questions that no one has posed. Michael Gazzaniga has advanced the notion of "interpreter" as a way of explaining the generation of consciousness. Moreover, he has related it, quite sensibly, to the machinery of the left hemisphere and to the language processes therein. I like his idea very much (in fact, there is a distinct ring of truth to it), but I believe it applies fully only to the level of the autobiographical self and not quite to that of the core self.[9]

In brains endowed with abundant memory, language, and reasoning, narratives with this same simple origin and contour are enriched and allowed to display even more knowledge, thus producing a well-defined protagonist, an autobiographical self. Inferences can be added, and actual interpretations of the proceedings can be produced. Still, as we shall see in the next chapter, the autobiographical self can be constructed only by means of the core self mechanism. The core self mechanism as just described, anchored in the protoself and its primordial feelings, is the central mechanism for the production of conscious minds. The complex devices required to extend the process to the autobiographical self level are dependent on the normal operation of the core self mechanism.

Would the mechanism for connecting self and object apply only to actually perceived objects and not to recalled objects? It would not. Given that when we learn about an object, we make records not just of its appearance but also of our interactions with it (our eye and head movements, our hand movements, and so forth), recalling an object encompasses recalling a varied package of memorized motor interactions. As in the case of actual motor interactions with an object, recalled or imaginary motor interactions can modify the protoself instantly.

Should this idea be correct, it would explain why we do not lose consciousness when we daydream in a silent room with our eyes closed—a rather comforting thought, I guess.

In conclusion, the production of pulses of core self relative to a large number of objects interacting with the organism guarantees the production of object-related feelings. In turn, such feelings construct a robust self process that contributes to the maintenance of wakefulness. The core self pulses also confer degrees of value upon the images of the causative object, thus giving it more or less salience. This differentiation of the flowing images organizes the landscape of the mind, shaping it in relation to the needs and goals of the organism.

The Core Self State

How might the brain implement the core self state? The search takes us first to fairly local processes, involving a limited number of brain regions, and then to brain-wide processes, involving many regions simultaneously. The steps related to the protoself are not difficult to conceive neurally. The interoceptive component of the protoself is based in the upper brain stem and in the insula; the sensory portal component is based in the conventional somatosensory cortices and frontal eye fields.

The status of some of these components has to change for the core self to emerge. We have seen that when a perceived object precipitates an emotional reaction and alters the master interoceptive maps, a modification of the protoself ensues, thus altering the primordial feelings. Likewise, the sensory portal components of the protoself change when an object engages a perceptual system. As a consequence, the regions involved in making images of the body are inevitably changed at protoself sites—brain stem, insular cortex, and somatosensory cortices. These varied events generate microsequences of images that are introduced into the mind process, by which I mean that they are introduced

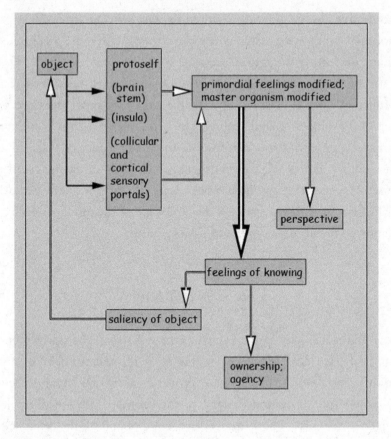

Figure 8.4: Schematic of core self mechanisms. The core self state is a composite. The main components are *feelings of knowing* and the *saliency* of the object. Other important components are *perspective* and the sense of *ownership* and *agency*.

into the image workspace of the early sensory cortices and of select regions of the brain stem, those in which feeling states are generated and modified. The microsequences of images succeed each other like beats in a pulse, irregularly but dependably, for as long as events continue to happen and the wakefulness level is maintained above threshold.

Up to this point, in the simplest instances of core self state, there probably is no need for a central coordination device and no need at all for a single screen to display the images. The chips (the images) fall where they must (the image-making regions) and enter the mind stream as they do, in their appropriate time and order.

For the construction of the self state to be complete, however, the modified protoself must be connected with the images of the causative object. How might that happen? And how does the ensemble of these disparate sets of images get to be organized so that it constitutes a coherent scene and thus a fully fledged pulse of core self?

Timing is likely to play a role here too, when the causative object begins to be processed and changes in the protoself begin to occur. These steps take place in close temporal proximity, in the form of a narrative sequence imposed by real-time occurrences. The first level of connection between modified protoself and object would emerge naturally out of the time sequence with which the respective images are generated and incorporated into the cortege of the mind. In brief, the protoself needs to be open for business—awake enough to produce the primordial feeling of existence born out of its dialogue with the body. Then the processing of the object has to modify the varied aspects of the protoself, and these events have to be connected to each other.

Might there be a need for neural coordinating devices to create the coherent narrative that defines the protoself? The answer depends on how complex the scene is and whether it involves multiple objects. When it does involve multiple objects, and even if the complexity is nowhere near the level that we shall consider in the next chapter regarding the autobiographical self, I believe we do need coordinating devices to achieve coherence. There are good candidates for that role, located at the subcortical level.

The first candidate is the superior colliculus. Its candidacy will evoke smiles, even if the coordinating credentials of this tried-and-true device cannot be questioned. For reasons outlined in Chapter 3, the deep layers of the superior colliculi are suited to this role. By offering the possibility of making superpositions of images of different aspects of the internal and external worlds, the deep layers of the colliculi are a model of what the mind-making and self-making brain eventually became.[10] The limitations are obvious, however. We cannot expect the colliculi to be the lead coordinator of cortical images when it comes to the complexity of the autobiographical self.

The second candidate for the role of coordinator is the thalamus, specifically the associative nuclei of the thalamus, whose situation is ideal to establish functional linkages among separate sets of cortical activity.

Touring the Brain as It Constructs a Conscious Mind

Imagine the following setting: I am watching pelicans feed breakfast to their young. They fly gracefully over the ocean, sometimes barely above the surface, sometimes higher up. When they spot a fish, they suddenly plunge toward the ocean surface, their Concorde-like beaks in landing attitude, their wings pulled back in a beautiful delta shape. They disappear into the water to emerge a second later, triumphant, with a fish.

My eyes are busy following the pelicans; as the pelicans move about, nearer or farther away, the lenses in my eyes modify their focal distance, the pupils adjust to the varying light, and the eye muscles work briskly to follow the birds' swift movements; my neck helps with appropriate adjustments, and my curiosity and interest are positively rewarded by observing such a remarkable ritual; I am enjoying the show.

As a result of all this real-life bustle and brain bustle, signals are arriving in my visual cortices, fresh from retinal maps that plot the pelicans and define their appearance as the object-to-be-known. A profusion of moving images is being made. On parallel tracks, signals are also being processed in a variety of brain regions: in the frontal eye fields (area 8, which is concerned with eye movements but not with visual images per se); in the lateral somatosensory cortices (which plot the muscular activity of the head, neck, and face); in emotion-related structures in the brain stem, the basal forebrain, the basal ganglia, and the insular cortices (whose combined activities help generate my pleasant feelings about the scene); in the superior colliculi (whose maps are receiving information about the visual scene, eye movements, and state of the body); and in associative nuclei of the thalamus engaged by all the signal traffic in the cortex and brain-stem regions.

And what is the upshot of all these changes? The maps that plot the state of sensory portals and the maps that pertain to the interior state of the organism are registering a perturbation. A modification of the protoself's primordial feeling now becomes differential feelings of knowing relative to the engaging objects. As a result, the recent visual maps of the object-to-be-known (the feeding pelican flock) are made more salient than other material being processed nonconsciously in my mind. That other material might compete for conscious treatment, but it does not succeed because, for a variety of reasons, the pelicans are so interesting to me, meaning valuable. Reward nuclei in regions such as the brain stem's ventral tegmental area, the nucleus accumbens, and the basal ganglia accomplish the special treatment of the pelican images by selectively releasing neuromodulators in image-making areas. A sense of ownership of the images, as well as a sense of agency, arises from such feelings of knowing. At the same time, the changes in the sensory portals have placed the object-to-be-known in a definite perspective relative to me.[11]

Out of this global-scale brain map, core self states emerge in pulse-like fashion. But suddenly the phone rings, and the spell is broken. My head and eyes move reluctantly but inexorably to the receiver. I get up. And the whole cycle of conscious mind-making starts anew, now focused on the telephone. The pelicans are gone from my sight and from my mind; the telephone is in.

9

The Autobiographical Self

Memory Made Conscious

Autobiographies are made of personal memories, the sum total of our life experiences, including the experiences of the plans we have made for the future, specific or vague. Autobiographical selves are autobiographies made conscious. They draw on the entire compass of our memorized history, recent as well as remote. The social experiences of which we were a part, or wish we were, are included in that history, and so are memories that describe the most refined among our emotional experiences, namely, those that might qualify as spiritual.

While the core self pulses away relentlessly, always "online," from hint half-hinted to blatant presence, the autobiographical self leads a double life. On the one hand, it can be overt, making up the conscious mind at its grandest and most human; on the other, it can lie dormant, its myriad components waiting their turn to become active. That other life of the autobiographical self takes place offscreen, away from accessible consciousness, and that is possibly where and when the self matures, thanks to the gradual sedimentation and reworking of one's memory. As lived experiences are reconstructed and replayed, whether in conscious reflection or in nonconscious processing, their substance is reassessed and inevitably rearranged, modified minimally or very much in terms of their factual composition and emotional accompaniment. Entities and events acquire new emotional weights during this process. Some frames of the recollection are dropped on the mind's cutting-room floor, others are restored and enhanced, and others still are so deftly combined either by our wants or by the vagaries of chance that they create new scenes that were never shot. That is how, as years pass, our own history is subtly rewritten. That is why facts can acquire a new significance and why the music of memory plays differently today than it did last year.

Neurologically speaking, this building and rebuilding job occurs largely in nonconscious processing, and for all we know, it may even occur in dreams, although it can emerge in consciousness on occasion. It makes use of the convergence-divergence architecture to turn the encrypted knowledge contained in dispositional space into explicit, decrypted displays in the image space.

Fortunately, given the abundance of records of one's lived past and anticipated future, we do not need to recall all of them or even most of them, whenever our selves operate in autobiographical mode. Not even Proust would have needed to draw on all of his richly detailed and long-ago past to construct a moment of full-fledged self-Proustiness. Thankfully, we rely on key episodes, a collection of them actually, and, depending on the needs of the moment, we simply recall a certain number of them and bring them to bear on the new episode. In certain situations, the number of summoned episodes can be very high, a true flood of memories suffused with the emotions and feelings that first went with them. (One can always count on Bach to bring about such a situation.) But even when the number of episodes is limited, the complexity of memoranda involved in structuring the self is, to put it modestly, very large. Therein lies the problem of constructing the autobiographical self.

Constructing the Autobiographical Self

I suspect that the brain's strategy for constructing the autobiographical self is as follows. First, substantial sets of defining biographical memories must be grouped together so that each can be readily treated as an individual object. Each such object is allowed to modify the protoself and produce its pulse of core self, with the respective feelings of knowing and consequent object saliency in tow. Second, because the objects in our biographies are so numerous, the brain needs devices capable of coordinating the evocation of memories, delivering them to the protoself for the requisite interaction, and holding the results of the interaction in a coherent pattern connected to the causative objects. This is not a trivial problem. In effect, complex levels of autobiographical self—those that, for example, include substantial social aspects—encompass so many biographical objects that they require numerous core self pulses. As a consequence, constructing the autobiographical self demands a neural apparatus capable of obtaining multiple core self pulses, within a brief time window, for a substantial number of components and holding the results together transiently, to boot.

From a neural standpoint the coordinating process is especially complicated by the fact that the images that constitute an autobiography are largely implemented in the image workspaces of the cerebral cortex, based on recall from dispositional cortices, and yet, in order to be made conscious, those same images need to interact with the protoself machinery, which, as we have seen, is largely located at brain-stem level. Constructing an autobiographical self calls for very elaborate coordinating mechanisms, something that the construction of the core self can, by and large, dispense with.

By way of a working hypothesis, then, we can say that constructing the autobiographical self depends on two conjoined mechanisms. The first is subsidiary to the core self mechanism and guarantees that each biographical set of memories is treated as an object and made conscious

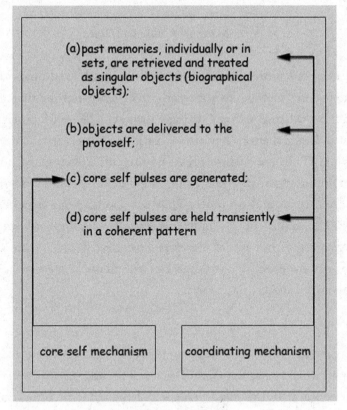

(a) past memories, individually or in sets, are retrieved and treated as singular objects (biographical objects);

(b) objects are delivered to the protoself;

(c) core self pulses are generated;

(d) core self pulses are held transiently in a coherent pattern

core self mechanism

coordinating mechanism

Figure 9.1: The autobiographical self: neural mechanisms.

in a core self pulse. The second accomplishes a brain-wide operation of coordination that includes the following steps: (1) certain contents are evoked from memory and displayed as images; (2) the images are allowed to interact in an orderly manner with another system elsewhere in the brain, namely, the protoself; and (3) the results of the interaction are held coherently during a certain window of time.

The structures involved in constructing the autobiographical self include all those required for the core self, in the brain stem, thalamus, and cerebral cortex, and, in addition, the structures involved in the coordination mechanisms discussed below.

The Issue of Coordination

Before I say one more word about coordination, I would like to make certain that my idea is not misinterpreted. The coordinating devices that I am postulating are *not* Cartesian theaters. (There is no play being performed inside them.) They are *not* consciousness centers. (There is no such thing.) They are *not* interpreter homunculi. (They know nothing, they do not interpret anything.) They are precisely what I am hypothesizing them to be and no more. They are spontaneous *organizers* of a process. The results of the entire operation *materialize not within the coordinating devices* but rather *elsewhere,* specifically, within the image-making, mind-generating structures of the brain located in both the cerebral cortex and the brain stem.

The coordination is driven not by some mysterious agent external to the brain but rather by natural factors such as the order of introduction of imaged contents in the mind process and the value accorded to those contents. How is the valuation achieved? Consider that any image being processed by the brain is automatically appraised and marked with a value in a process based on the brain's original dispositions (its biological value system), as well as on the dispositions acquired over lifelong learning. The marking stamp is added during the original perception and is recorded along with the image, but it is also revived during every instance of recall. In brief, confronted with certain sequences of events and a wealth of past knowledge filtered and marked by value, the brain's coordinating devices assist with the organization of the current contents. Moreover, the coordinating devices deliver the images to the protoself system and finally hold the results of the interaction (pulses of core self) in a transient coherent pattern.

The Coordinators

In the working hypothesis presented here, the first stage of the implementation of the neural autobiographical self requires structures and mechanisms already discussed for the core self. But there is something distinctive about the structures and mechanisms needed to implement the second stage of the process, namely, the brain-wide coordination described earlier.

What are the candidates for this large-scale system-coordination role? Several possible structures come to mind, but only a few can be seriously considered. An important candidate is the thalamus, a perpetual presence in any discussion of the neural basis of consciousness, specifically its collection of associative nuclei. The intermediate position of the thalamic nuclei, between the cerebral cortex and the brain stem, is ideal for signal brokering and coordination. Although the associative thalamus is busy enough constructing the background fabric of any image, it plays a very important, albeit perhaps not the lead, role when it comes to coordinating the contents that define the autobiographical self. I will say more about the thalamus and coordination in the next chapter.

What are the other likely candidates? A strong contender is a composite collection of regions in both cerebral hemispheres that is distinguished by its connectional architecture. Each region is a macroscopic node located at a major crossroads of convergent and divergent signaling. I described them as convergence-divergence regions or CDRegions in Chapter 6 and indicated that they are made of numerous convergence-divergence zones. CDRegions are strategically located within high-order association cortices but not within the image-making sensory cortices. They surface in sites such as the temporoparietal junction, the lateral and medial temporal cortices, the lateral parietal cortices, the lat-

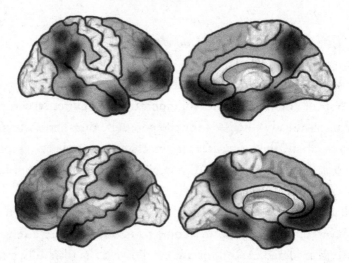

Figure 9.2: The task of coordinating the varied images generated by ongoing perception and recall is assisted by convergence-divergence regions (CDRegions), which are located within the nonmapped association cortices. The approximate location of the main CDRegions is suggested in the diagram (darkly shaded areas): the polar and medial temporal cortices, the medial prefrontal cortices, the temporoparietal junctions, and the posteromedial cortices (PMCs). In all likelihood, there are other such regions. Most of the CDRegions depicted in the figure are also part of Raichle's "default network" discussed later in this chapter. See Chapter 6 and Figures 6.1 and 6.2 for the architecture of these regions. See Figure 9.4 for connectional details of one CDRegion, the PMCs.

eral and medial frontal cortices, and the posteromedial cortices. These CDRegions hold records of previously acquired knowledge regarding the most diverse themes. The activation of any of these regions promotes the reconstruction, by means of divergence and retroactivation into image-making areas, of varied aspects of past knowledge, including those that pertain to one's biography, as well as those that describe genetic, nonpersonal knowledge.

Conceivably, the main CDRegions could be further integrated by long-range cortico-cortical connections of the kind first identified by

Jules Déjérine a century ago. Such connections would introduce yet another level of interareal coordination.

One of the main CDRegions, the posteromedial cortices (PMCs), appears to have a higher functional hierarchy relative to the others and exhibits several anatomical and functional traits that distinguish it from the rest. A decade ago I suggested that the PMC region was linked to the self process, albeit not in the role I now envision. Evidence obtained in recent years suggests that the PMC region is indeed involved in consciousness, quite specifically in self-related processes, and has provided previously unavailable information regarding the neuroanatomy and physiology of the region. (The evidence is discussed in the last sections of this chapter.)

The final candidate is a dark horse, a mysterious structure known as the claustrum, which is closely related to the CDRegions. The claustrum, which is located between the insular cortex and the basal ganglia of each hemisphere, has cortical connections that might potentially play a coordinating role. Francis Crick was convinced that the claustrum was a sort of director of sensory operations charged with binding disparate components of a multisensory percept. The evidence from experimental neuroanatomy does reveal connections to varied sensory regions, thus making the coordinating role quite plausible. Intriguingly, it has a robust projection to the important CDRegion that I mentioned earlier, the PMC. The discovery of this strong link occurred only after Crick's death and was thus not included in the posthumously published article that he wrote with Christof Koch, in which he made his case.[1] The problem with the claustrum's candidacy as coordinator resides in its small scale when we consider the job that needs to be performed. On the other hand, given that we should not expect any of the structures discussed earlier to perform the coordinating job single-handedly, there is no reason why the claustrum should not make a relevant contribution to the construction of the autobiographical self.

A Possible Role for the Posteromedial Cortices

We need additional research to determine the specific role the PMCs play in the construction of consciousness. Later in this chapter, I review evidence from varied sources: anesthesia research, sleep research, research on neurological conditions (ranging from coma and vegetative state to Alzheimer's disease), and functional neuroimaging studies of self-related processes. But first let's look at the PMC evidence that appears most solid and interpretable—evidence from experimental neuroanatomy. I'll speculate on the possible workings of the PMCs and on the reasons why they should be investigated.

When I proposed that the PMCs would play a role in generating subjectivity, there were two strands of thinking behind the idea. One strand concerned the behavior and presumed mental status of neurological patients with focal damage to this region, which includes the damage caused by late-stage Alzheimer's disease, as well as extremely rare strokes and brain metastases from cancer. The other strand related to a theoretical search for a brain region physiologically suitable to bring together information about both the organism *and* the objects and events with which the organism interacts. The PMC region was one of my candidates, given that it appeared to be located at an intersection of pathways associated with information from the visceral interior (interoceptive), from the musculoskeletal system (proprioceptive and kinesthetic), and from the outside world (exteroceptive). The factual strands are not in question, but I no longer see a need for the functional role I had envisioned. Still, the hypothesis prompted investigations that yielded important new information.

Making headway with the hypothesis was not easy; the main problem was that the neuroanatomical information available on this region was quite limited. Some valuable studies had begun to chart the connectivity of parts of the PMC,[2] but the overall wiring diagram of the region had not been investigated. In fact, the region was known not by an umbrella term but rather by its component parts, namely, the posterior cingulate cortex, the retrosplenial cortex, and the precuneus. The PMCs, by whatever name, were definitely not yet on the radar of notable brain areas.

In order to explore the hypothesis that the PMC was involved in consciousness, it was necessary to acquire previously unavailable knowledge about the connectional neuroanatomy of the PMCs. For this reason, our research group undertook an experimental neuro-anatomical study in nonhuman primates. The experiments were conducted in Josef Parvizi's laboratory in collaboration with Gary Van Hoesen. In essence the study consisted of making, in experimental macaque monkeys, numerous injections of biological tracers into all the territories whose neural connectivity we needed to investigate. Once injected into a given brain region, biological tracers are absorbed by individual neurons and transported along their axons all the way to their natural destinations, whatever the neurons are currently connecting to. These are the so-called anterograde tracers. Another kind of biological tracer, the retrograde kind, is taken up by axon terminals and transported in reverse, from wherever the terminals are, back to the cell bodies of the neurons, at their points of origin. The upshot of all the tracer travels is the possibility of charting, for each target region, the sites of origin of the connections the region receives, as well as the sites toward which the region sends its messages.

The PMCs are constituted by several subregions. (In Brodmann's cytoarchitectonic map, they are areas 23a/b, 29, 30, 31, and 7m.) The interconnectivity of these subregions is so intricate that it is reasonable, to some degree, to treat them as a functional unit. Some distinct connectional affiliations within the subsectors open the possibility that some of them may have distinct functional roles to play. The umbrella term we coined for the ensemble appears justified, at least for the time being.

. . .

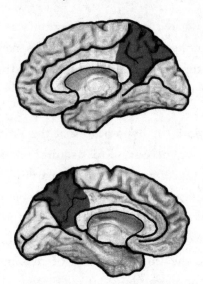

Figure 9.3: The location of the posteromedial cortices
in the human brain.

The pattern of PMC connections, as reported in the first publication to
come from these laborious and time-consuming investigations,[3] is sum-
marized in Figure 9.4. It can be described as follows:

1. Inputs from parietal and temporal association cortices,
entorhinal cortices, and frontal cortices converge in the PMCs, as
do inputs from the anterior cingulate cortex (a principal recipient
of projections from the insula), the claustrum, the basal forebrain, the amygdala, the premotor region, and the frontal eye
fields. Thalamic nuclei, both intralaminar and dorsal, also project
to the PMCs.

2. With few exceptions, the sites that originate converging
inputs to the PMCs also receive diverging outputs from them,
exceptions being the ventromedial prefrontal cortex, the claustrum, and the intralaminar nuclei of the thalamus. Some sites
that do not project to the PMCs do receive PMC projections,
namely the caudate and putamen, the nucleus accumbens, and
the periaqueductal gray.

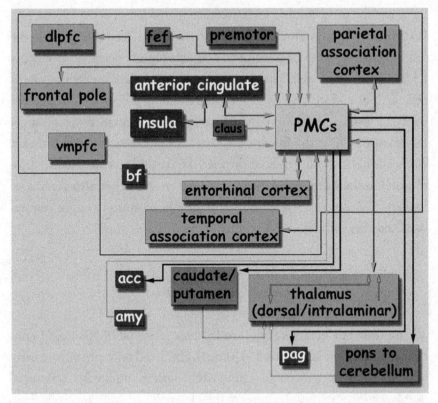

Figure 9.4: The pattern of neural connections to and from the postero-medial cortices (PMCs), as determined in a study conducted in the monkey. Abbreviations: dlpfc = dorsolateral prefrontal cortex; fef = frontal eye fields; vmpfc = ventromedial prefrontal cortex; bf = basal forebrain; claus = claustrum; acc = nucleus accumbens; amy = amygdala; pag = peri-aqueductal gray.

3. There are no connections to or from the PMCs relative to the early sensory cortices or the primary motor cortices.

4. From the results described under 1 and 2, it is apparent that the PMCs are a high-level convergence and divergence region. It is a prominent member of the club of CDRegions that I regard as good candidates for coordinating the contents in the conscious mind, and it even has an important connection with another potential coordinator, the claustrum, which significantly projects to the PMCs but is poorly reciprocated.

A recent study conducted in humans has added support for the idea that the PMCs are neuroanatomically distinct.[4] The study, which was led by Olaf Sporns, used a modern technique of magnetic resonance imaging, diffusion spectrum imaging, which produces images of neural connections and of their approximate spatial distribution. The authors used their imaging data to construct maps of the connectional arrangements throughout the human cerebral cortex. They identified several connectional hubs throughout the cerebral cortex, several of which correspond to the CDRegions I have been discussing. They also concluded that the PMC region constitutes a unique hub, more strongly interrelated to other hubs than any of the others.

The PMCs at Work

We are now in a better position to imagine how the PMCs might contribute to the conscious mind. Although this is a sizable portion of cerebral cortex, the power of the PMCs resides not with territorial possessions but with the company they keep. The PMCs receive signals from most high-order sensory association regions and premotor regions and largely return the favors. Brain areas rich in convergence-divergence zones, which hold the key to composites of multimodal information, are thus able to signal to the PMCs and by and large can be signaled back. The PMCs also receive signals from subcortical nuclei involved in wakefulness and in turn signal to a variety of subcortical regions related to attention and reward (in the brain stem and basal forebrain), as well as to regions capable of producing motor routines (such as the basal ganglia and the periaqueductal gray).

What are the received signals likely to be about, and what do the PMCs do with them? We do not know for certain, but the huge disproportion between the profusion and strength of the projections toward the PMCs and the actual territory in which they land suggests an answer. The PMCs are mostly of older vintage, territories that one thinks of as holding dispositions rather than explicit maps. The PMCs are not modern early sensory cortices like those of vision or hearing, where detailed maps of things and events can be assembled. Let us say that the PMC gallery has not enough wall space to exhibit large paintings or, for that matter, to present puppet shows. But that is just fine because the cortices that signal to the PMCs are not like early sensory cortices either; they cannot exhibit large paintings or present puppet shows any more than the PMCs can; they too are largely dispositional, convergence-divergence zone holders of recorded information.

Given their design, the PMCs as a whole and their component submodules are likely to behave as convergence-divergence regions themselves. I envision that the information held by the PMCs as well as by their partners can be played back only by signaling back into other CDRegions in the club, which in turn can signal to early sensory cortices. Those are the cortices where images can be made and displayed—that is, where large paintings can be shown and puppet shows presented. Relative to the other convergence-divergence regions that interconnect with them, the PMCs have a special hierarchical rank. The PMC region sits higher on the totem pole, capable of interactive signaling with the other CDRegions.

How, then, does the PMC assist consciousness? By contributing to the assembly of autobiographical self states. This is what I envision: separate sensory and motor activities related to personal experience would have been originally mapped in the appropriate brain regions, cortically and subcortically, and the data recorded in convergence-divergence zones and in convergence-divergence regions. In turn, the PMCs would have constituted a higher-order CDRegion record interconnected with the other CDRegions. The arrangement would allow activity in the PMCs to access larger, highly distributed data sets, but with the advantage that the access command would come from a relatively small and thus spatially manageable territory. The PMCs could support the establishment of momentary and temporally cohesive displays of knowledge.

If the PMCs' pattern of neuroanatomical connections is noteworthy, so is their anatomical location. The PMCs are located near the midline, the left set looking across the interhemispheric divide at the right set. This geographic position within the brain volume is convenient for both convergence and divergence connectivity relative to most regions of the cortical mantle, and it is ideal for receiving signals from the thalamus and reciprocating them. Curiously, the location also affords protection from external impact, and, because it is supplied by three major and separate blood vessels, it makes the PMCs relatively immune to the sort of vascular damage or trauma that could radically destroy them.

As I have previously emphasized, consciousness-related structures share several anatomical traits. First, either at the subcortical or the cortical level, they tend toward the old vintage. This should not be surprising given that the beginnings of consciousness occurred late in biological evolution but are not at all a recent evolutionary development. Second, both cortical and subcortical structures tend to be placed at or near the midline, and, just like the PMCs, they like to look at their twin siblings across the brain's midline—this is the case with thalamic and hypothalamic nuclei, as well as with brain-stem tegmental nuclei. Evolutionary age and convenience of location relative to widespread signal distribution are closely correlated here.

The PMCs would operate as a partner to the network of cortical CDRegions. But the role of the other CDRegions and the importance of the protoself system is such that consciousness is likely to be affected but not abolished following the hypothetical destruction of the entire PMC region, provided all the other CDRegions and the protoself system remain intact. Consciousness would be restored, albeit not at its peak. The situation of late-stage Alzheimer's disease, which I describe in the next section, is different in the sense that the PMC insult is virtually the last straw in a process of gradual ravage that has already disabled other CDRegions and the protoself system.

Other Considerations on the Posteromedial Cortices

ANESTHESIA RESEARCH

In some respects, general anesthesia is an ideal means to investigate the neurobiology of consciousness. It is one of the most spectacular developments of medicine and has saved the lives of millions of people who otherwise could not have had surgery. One often thinks of general anesthesia as a painkiller, since its effects preclude the pain that surgical wounds would cause, but the truth is that anesthesia precludes pain in the most radical way possible: it suspends consciousness altogether, not merely pain but all aspects of the conscious mind.

Superficial levels of anesthesia reduce consciousness lightly, leaving room for some unconscious learning and the occasional "breakthrough" of conscious processing. Deep levels of anesthesia cut deep into the conscious process and are, in point of fact, pharmacologically controlled variations on the vegetative state or even coma. That is what your surgeon needs if he is to work in peace inside your heart or your hip joint. You must be far, far away from it all, so deeply asleep that your muscular tone is as tough as jelly and you are not able to move. Stage III anesthesia is the ticket, and at that stage you will hear nothing, feel nothing, and think of nothing. When the surgeon talks to you, you will not respond.

The history of anesthesia has provided surgeons with numerous pharmacological agents to work with, and the search for the molecules that can do the most efficient job with minimal risks and little toxicity is an ongoing effort. By and large, anesthetics do their job by increasing inhibition in neural circuits. This can be achieved by strengthening the action of GABA (gamma-aminobutyric acid), the leading inhibitory transmitter in the brain. Anesthetics act by hyperpolarizing neurons and blocking acetylcholine, an important molecule in normal neuron-to-neuron communication. It was commonly thought that anesthetic agents worked by depressing brain function across the board, bringing down the activity of neurons most everywhere. But recent studies have shown that some anesthetics work very selectively, exerting their action at specific brain sites. A case in point is propofol. As shown in functional neuroimaging studies, it does its splendid job by working principally at three sites: the posteromedial cortices, the thalamus, and the brain-stem tegmentum. While the relative importance of each site in the production of unconsciousness is unknown, the decreases in level of consciousness are correlated with the decrease of regional blood flow in the posteromedial cortices.[5] But the evidence goes well beyond propofol. Other anesthetic agents seem to have comparable effects, as a comprehensive review demonstrates. Three paramedian brain territories instrumental in building consciousness are selectively depressed by propofol anesthesia.

SLEEP RESEARCH

Sleep is a natural setting for the study of consciousness, and sleep studies were early contributors to the understanding of the problem. It has been well established that electroencephalographic rhythms, the distinct patterns of electrical activity generated by the brain, are associated with specific stages of sleep. It is notoriously difficult to peg the origin of electroencephalographic patterns to particular brain regions, and that is where the spatial localization of functional neuroimaging techniques has come in handy to complete the picture. Using imaging techniques, it has been possible, over the past decade, to take a closer look at specific brain regions during varied stages of sleep.

For example, consciousness is deeply depressed during slow-wave sleep, also known as non–rapid eye movement sleep or N-REM. This is the deep slumber of the kind and the just, the slumber from which only the unkind and unjust alarm clock will wake us up. This is "dreamless sleep," although the complete absence of dreams appears to apply only to the first part of the night. Functional neuroimaging studies show

that in slow-wave sleep, activity is reduced in a number of brain regions, most prominently in parts of the brain-stem tegmentum (at the
pons and midbrain), the diencephalon (the thalamus and the hypo-
thalamus/basal forebrain), the medial and lateral parts of the prefrontal
cortex, the anterior cingulate cortex, the lateral parietal cortex, and the
PMCs. The pattern of functional reduction in slow-wave sleep is less
selective than in general anesthesia (there is no reason why the pattern
should be the same), but as in anesthesia, it does not suggest an across-
the-board depression of function. The pattern does include, prominently, the three correlates of consciousness-making (brain stem,
thalamus, and PMCs), and it does show that all three are depressed.

In brief, the level of activity in the PMCs is highest during wakefulness and lowest during slow-wave sleep. During REM sleep the PMCs
operate at intermediate levels. This makes some sense. Consciousness is
mostly suspended during slow-wave sleep; in dream sleep, things do
happen to a "self." The dream self is not the normal self, of course, but
the brain state that goes with it appears to recruit the PMCs.

Consciousness is also depressed during rapid eye movement (REM)
sleep, during which dreams are most prevalent. But REM sleep allows
dream contents to enter consciousness, either via learning and subsequent recall or via so-called paradoxical consciousness. The brain
regions whose activity is most markedly decreased during REM are
the dorsolateral prefrontal cortex and the lateral parietal cortex; predictably, the decrease in activity of the PMCs is far less marked.[6]

In brief, the level of activity in the PMCs is highest during wakefulness and lowest during slow-wave sleep. During REM sleep the PMCs
operate at intermediate levels. This makes some sense. Consciousness is
mostly suspended during slow-wave sleep; in dream sleep, things do
happen to a "self." The dream self is not the normal self, of course, but
the brain state that goes with it appears to recruit the PMCs.

THE PMCS' INVOLVEMENT IN THE DEFAULT NETWORK

In a series of functional imaging studies using both positron-emission
tomography and functional magnetic resonance, Marcus Raichle called
attention to the fact that when subjects are at rest, not engaging in a task
requiring focused attention, a selective subset of brain regions appears
consistently active; when attention is directed to a specific task, the
activity of these regions decreases slightly, but never to the degree

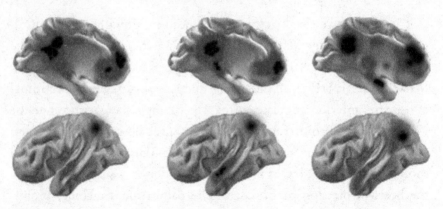

Figure 9.5: The PMCs, along with other CDRegions, are prominently activated in a variety of functional imaging tasks involving self-reference. Such tasks include recalling autobiographical memory, anticipating future events, and making moral judgments.

noted in anesthesia, for example.[7] The subset of regions includes the medial prefrontal cortex; the temporoparietal junction, structures in the medial and anterior temporal cortex, and the PMCs, all regions that we now know to be extensively interconnected. Most of the attention focused on the PMCs has actually come from their membership in this club of regions.

Raichle has suggested that the activity of this network represents a "default mode" of operation, a mode that is disrupted by tasks requiring externally directed attention. In tasks requiring internally directed and self-oriented attention, such as in the retrieval of autobiographical information and in certain emotional states, we and others have demonstrated that the decrease of activity in the PMCs is less pronounced or may fail to appear. In fact, in such conditions, there may be an actual increase.[8] Examples are the recall of autobiographical memories, the recall of plans made for a possible future, a number of theory-of-mind tasks, and a host of tasks that involve judgments of people or situations within a moral framework.[9] In all those tasks, there tends to be one more significant site of activity, albeit not as extensive: another medial territory, located anteriorly in the prefrontal cortex. We know that neuroanatomically this is also a convergence-divergence region.

. . .

Raichle has emphasized the intrinsic aspect of the default mode of operation and has related it, quite sensibly, to the very high energy consumption associated with intrinsic brain activity, as opposed to activity driven by external stimulation—in all probability, the PMCs are the most highly metabolic region of the entire cerebral cortex.[10] This too is compatible with the role I am proposing for the PMCs in consciousness, that of an important integrator/coordinator that would remain active at all times, attempting to hold highly disparate sets of background activity in a coherent pattern. How does the seesaw pattern of the default mode of operation fit with the idea that a region such as the PMCs would serve consciousness? It possibly reflects the background-foreground dance played by the self within the conscious mind. When we need to attend to external stimuli, our conscious mind brings the object under scrutiny into the foreground and lets the self retreat into the background. When we are unsolicited by the outside world, our self moves closer to center stage and may even move further forward when the object under scrutiny is our own person, alone or in its social setting.

RESEARCH ON NEUROLOGICAL CONDITIONS

The list of neurological conditions in which consciousness is compromised is mercifully short: coma and vegetative states, certain kinds of epileptic state, and the so-called akinetic mute states that may be caused by certain strokes, tumors, and late-stage Alzheimer's disease. In coma and vegetative states the compromise is radical, akin to a sledgehammer applied pointedly and unkindly to a brain territory.

Alzheimer's Disease. Alzheimer's, a uniquely human disease, is also one of the most serious health problems of modern times. As we attempt to understand it, however, and on a rather positive note, the condition has also become a source of valuable information about mind, behavior, and brain. The contributions of Alzheimer's disease to the understanding of consciousness are only now becoming apparent.

Beginning in the 1970s, I had the opportunity of following many patients with this condition and the privilege of studying their brains at postmortem, both the gross specimen and the microscopic material. In those years part of our research program was devoted to Alzheimer's disease, and my colleague and close collaborator Gary W. Van Hoesen was a leading expert in the neuroanatomy of the Alzheimer's brain. Our main goal then was understanding how circuit changes in the Alzheimer's brain could cause the disturbance of memory that charac- terized the condition.

Most patients with typical Alzheimer's disease do not have distur- bances of consciousness, either early in the disease or in its midstages. The first years of the disease are hallmarked by progressive defects in learning new factual information and in recalling previously learned factual information. Difficulties with judgment and spatial navigation are also common. Early on the touch of the disease may be so light that social graces are preserved and some semblance of life normality does persist for a while.

In the early 1980s our research group, which by then included Brad Hyman, established a reasonable cause for the factual memory defect in Alzheimer's disease: the extensive neuropathological changes in the entorhinal cortex and in the adjoining fields of the anterior temporal lobe cortices.[11] The hippocampus, the brain structure needed to lock in new memories of facts elsewhere in the brain, was effectively discon- nected from the entorhinal/anterior temporal lobe cortices. As a conse- quence, new facts could not be learned. In addition, as the disease progressed, the anterior temporal lobe cortices were themselves so damaged that they prevented access to unique, previously learned fac- tual information. In effect, the bedrock of autobiographical memory was eroded and was eventually just as wiped out as in patients with mas- sive destruction of the temporal lobe caused by herpes simplex enceph- alitis, a viral infection whose brunt also compromises the anterior temporal regions selectively. The cellular specificity of Alzheimer's dis- ease was uncanny. Most if not all neurons of layers II and IV of the entorhinal cortex were turned into tombstones, the best description for what is left of neurons after the disease changes them into neurofibrillary tangles. What this selective insult accomplished was a razor-sharp cut in the input lines to the hippocampus, which use layer II as a relay. And in order to make the severance complete, the insult also made an equally sharp cut in output lines from the hippocampus, those that use layer IV. Little wonder that factual memory is devastated in Alzheimer's.

As the disease progresses, however, along with other selective disturbances of mind, the integrity of consciousness begins to suffer. At first, the problem is predictably confined to autobiographical consciousness. Because memory about past personal events cannot be properly retrieved, the link between current events and the lived past becomes inefficient. Reflective consciousness in deliberative, offline processing is compromised. In all likelihood, part of this disturbance, though perhaps not all, is still due to medial temporal lobe dysfunction.

Further along its inexorable march, the ravage extends well beyond autobiographical processes. In the late stages of Alzheimer's, in those patients who receive good medical and nursing care and who survive the longest, a virtually vegetative state gradually sets in. The patients' connection to the world is reduced to a point where they resemble individuals in akinetic mutism. The patients initiate fewer and fewer interactions with the physical and human surroundings and respond to fewer and fewer prompts. Their emotions are muted. Their behavior is dominated by an absent, listless, vacant, unfocused, silent look.

What might account for the last turn of events in Alzheimer's disease? A definite answer is not possible because, along the years of disease, there are several sites of pathology in the Alzheimer brain and the pathology is not confined to neurofibrillary tangles. But to some extent the damage remains selective. The image-making sections of the brain, namely, the early sensory cortices of vision and hearing, are not burdened by

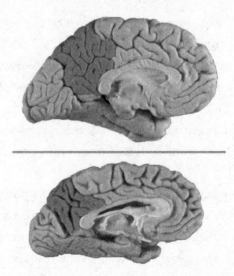

Figure 9.6: The top panel shows the medial view of the left cerebral hemisphere in a normal older individual. The PMC region is shaded. The bottom panel shows the same view in an individual of approximately the same age who had advanced Alzheimer's disease. The shaded PMC region is severely atrophic.

disease, nor are the movement-related regions in the cerebral cortex, the basal ganglia, and the cerebellum. On the other hand, some of the regions related to life regulation, on which the protoself depends, are progressively damaged. They include not just the insular cortex but also the parabrachial nucleus, something our group was also able to establish.[12] Finally, other brain sectors rich in CDRegions show severe damage. The PMCs figure prominently among the last.

The reason I am paying special attention to these facts is that early in the disease the PMCs show mostly neuritic plaques, but late in the disease the pathology is dominated by deposition of neurofibrillary tangles, the tombstones of former healthy neurons to which I alluded earlier. Their massive presence in the PMCs suggests that the operation of the region is severely compromised.[13]

We had been quite aware of important pathological changes in the PMCs, to which, in those days, we simply referred as the "posterior cingulate cortex and surround." But the repeated clinical observation of impaired consciousness in late-stage Alzheimer's, in cases of focal damage to this region, and its peculiar anatomical placement made me wonder if severely damaged PMCs might be the drop that made the cup overflow.[14]

Why is this region a target of Alzheimer's pathology? The reason may well be the same one my colleagues and I invoked, many years ago, to account for the prevalent pathological involvements of the medial temporal lobe regions in the same disease.[15] In normal health the entorhinal cortex and the hippocampus never cease their operations. They work day and night to assist with the processing of factual memories by initiating and consolidating memory records. Accordingly, local cellular toxicity associated with major wear and tear would take their toll on the region's precious neurons. Much the same reasoning would apply to the PMCs, given their nearly continuous operation in a variety of self-related processes.[16]

In sum, patients in late-stage Alzheimer's disease with evident compromises of consciousness have disproportionate neuronal damage and thus dysfunction of two brain territories whose integrity is necessary for normal consciousness: the PMCs and the brain-stem tegmentum. One should be prudent regarding the interpretation of these facts, given that there are other sites of dysfunction in Alzheimer's disease. At the same time it would be foolish not to consider this evidence.

And what about the patients themselves, who at this late stage in the disease receive yet another blow to the health of their brains? In the past my view has been, and still is today, that much as the new insult is painful to watch by those who are close to the victims, it is probably a blessing in disguise for the patient. Patients in this late stage and with this degree of impaired consciousness cannot possibly be aware of the ravages of the disease. They are shells of the human beings they once were, deserving of our love and care to the bitter end but now thankfully freed, to some degree, of the laws of pain and suffering that still apply to those who watch them.

COMA, VEGETATIVE STATE, AND THE CONTRAST
TO LOCKED-IN SYNDROME

Patients in coma are largely unresponsive to communication from the outside world, deep in a sleep in which even the pattern of breathing often sounds abnormal. They do not make meaningful gestures or utter meaningful sounds, let alone use words. None of the critical components of consciousness that I listed in Chapter 8 is in evidence. Wakefulness is gone for certain; and, based on the observable behavior, mind and self are reasonably presumed to be absent.

Patients in coma often have damage to the brain stem, and sometimes the damage trespasses into the hypothalamus. Most commonly this is caused by a stroke. We know that the damage must be located in the back part of the brain stem, the tegmentum, and more specifically in its upper tier. The upper tier of the tegmentum houses nuclei involved in life regulation but not those that are indispensable to maintaining breathing and cardiac function. In other words, when the damage involves the lower tier of the tegmentum as well, the result is death, not coma.

When the damage occurs in the front part of the brain stem, the result is also not coma but rather *locked-in syndrome,* a horrible condition in which the patient is entirely conscious but almost completely paralyzed. The patient can communicate only through blinking, sometimes with only one eye, sometimes through the upward movement of one eye. Yet they *see* perfectly well whatever is brought in front of their eyes and thus can read. They can hear perfectly well too and appreciate the world in fine detail. Their prison is nearly complete; only a dulling of the background emotional reactions somehow transforms a terrifying situation into a painful but barely tolerable one.

We know about those patients' unique experiences from a few dictated reports that some intelligent and observant patients had the courage to pursue, with expert help. The reports were not really dictated but rather "blinked," one letter a blink. I used to think that Lou

Gehrig's disease (amyotrophic lateral sclerosis) was the cruelest of neurological diseases. In Lou Gehrig's disease, which is a degenerative brain condition, equally conscious patients gradually lose the ability to move, to speak, and eventually to swallow. But once I saw my first patient with locked-in syndrome, I realized that it manages to be worse. The two best books by locked-in patients are small and simple but humanly rich. One of them, by Jean-Dominique Bauby, was turned into a surprisingly accurate movie, *The Diving Bell and the Butterfly,* directed by the painter Julian Schnabel. It offers nonspecialists a satisfactory documentary of the condition.[17]

Coma often transitions into a somewhat milder condition called vegetative state. The patient is still unconscious, but as previously noted the condition differs from coma on two counts. First, patients have sleep-wakefulness alternations, and when sleep or wakefulness occurs, its signature electroencephalographic pattern is present too. The patients' eyes can be open during the awake part of the cycle. Second, the patients do produce some movements and may respond with movements. But they do not respond with speech, and the movements they execute have no specificity. Vegetative state can transition into recovery of consciousness or remain stable, in which case it is called persistent vegetative state. In addition to damage to the brain-stem tegmentum and hypothalamus, which is the typical pathology of coma, vegetative state can result from damage to the thalamus and even from widespread damage to the cerebral cortex or underlying white matter.

How do coma and vegetative state relate to the PMC role, given that the causative lesions are located elsewhere? That question has been addressed in functional imaging studies aimed at investigating how generalized or restricted the functional changes are in the brain of patients in those conditions. The usual suspects do turn up, as revealed by major reductions of the function of the brain stem, the thalamus, and the PMCs, but the reduction of local metabolic rate for glucose observable at the PMCs is especially pronounced.[18]

But there is another relevant finding to report. Patients in a coma

commonly either die or improve quite modestly into a persistent vegetative state. Some patients, however, are more fortunate. They emerge
gradually from their state of profoundly impaired consciousness, and, as
they do, the most significant changes in brain metabolism occur at the
PMCs.[19] This suggests that the level of activity in this area is well correlated with the level of consciousness. Given that the PMCs are so highly
metabolic, one might be tempted to dismiss this finding as the result of
wholesale improvement of brain activity. The PMCs would improve
first merely because of their high metabolism. But that would not
explain why consciousness is regained at the same time.

A Closing Note on the Pathologies of Consciousness

The pathologies of consciousness have provided important pointers in
the delineation of a neuroanatomy of consciousness, and they have suggested aspects of the mechanisms proposed for the construction of core
and autobiographical selves. Perhaps it is helpful to close by establishing
a transparent link between human pathology and the hypotheses presented earlier.

Leaving aside the alterations of consciousness that arise naturally
from sleep or are induced by anesthetics under medical control, most
disturbances of consciousness result from profound brain dysfunction
of one sort or another. In some instances, the mechanism is chemical;
this is the case with overdoses of various drugs, including insulin given
for the treatment of diabetes, as well as with excessive blood levels of
glucose in untreated diabetes. The effect of these chemical molecules is
both selective and generalized. Given prompt and adequate treatment,
however, the conditions are reversible. On the other hand, structural
damage caused by head trauma, stroke, or certain degenerative diseases
often produces disturbances of consciousness from which complete
recovery is unlikely. Moreover, in some situations, brain damage can
also lead to seizures, during or after which altered states of consciousness are a prominent symptom.

Cases of coma and of vegetative state due to brain-stem damage compromise both the core self and the autobiographical self. In essence, the main protoself structures are either destroyed or severely damaged, and neither primordial feelings nor "feelings of what happens" can be generated. An intact thalamus and an intact cerebral cortex are not sufficient to compensate for the collapse of the core self system. Such conditions testify to the hierarchical precedence of the core self system and to the entire dependence of the autobiographical self system on that of the core self. This is important to note since the inverse is not true: the autobiographical self can be compromised in the presence of an otherwise intact core self.

Cases of coma or persistent vegetative state in which the brunt of the damage, rather than affecting the brain stem, compromises the cortex, the thalamus, or the connection of these structures to the brain stem may render the core self dysfunctional rather than destroy it, thus explaining the progression of some of these cases toward "minimal" consciousness and the recovery of some nonconscious mind-related activities. Cases of akinetic mutism and postseizure epileptic automatism cause reversible compromises of the core self system and a consequent alteration of the autobiographical self system. Some appropriate behaviors are present and, albeit automatic, suggest that mental processes are by no means abolished.

When autobiographical self disturbances appear independently, with an otherwise intact core self system, the cause is some aspect of memory dysfunction, an acquired amnesia. The most important cause of amnesia is the condition just discussed, Alzheimer's; other causes include viral encephalitis and acute anoxia (loss of brain oxygenation) as can occur in cases of cardiac arrest. In cases of amnesia, there is a considerable disruption of the unique memories that correspond to one's past and one's plans for the future. Obviously, patients with damage to both hippocampal-entorhinal regions, whose ability to make new memories is compromised, suffer from a progressive loss of scope in their autobiographical self because the new events of their lives are not properly recorded and integrated into their biographies. More serious is the situation of patients whose brain damage encompasses not only the hippocampal-entorhinal regions but also the regions around and beyond the entorhinal cortices, in the anterior sector of the temporal lobe. Such patients appear to be entirely conscious—their core self operations are intact—so much so that they are even conscious of their failures of recall. However the degree to which they can evoke their biographies, along with all the social information they carry, is diminished to a smaller or greater extent. The material with which an autobiographical self can be assembled is impoverished, either because it cannot be brought out of past records or because whatever is brought out cannot be properly coordinated and delivered to the protoself system, or perhaps both. The extreme case is that of patient B, whose biographical recall is largely confined to his childhood and is quite schematic. He knows that he was married and is the father of two sons, but he knows almost nothing concrete about his family members, whom he cannot recognize, in photographs or in person. His autobiographical self is severely compromised. On the other hand, another well-known amnesic patient, Clive Wearing, has far more preserved recall of his biography. He has not only a normal core self but a robust autobiographical self. A passage from a letter his wife, Deborah Wearing, wrote to me, explains why I think so:

> He can describe the approximate layout of his childhood bedroom, he knows that he sang in Erdington Parish Choir from an early age, he says he remembers being in the bomb shelter during the war and the sound of bombs in Birmingham. He knows a number of stubs of facts about his childhood and about his parents and siblings, he can sketch his adult autobiography—his Cambridge college where he was choral scholar; where he worked; the London Sinfonietta, the BBC Music Department, his career as a conductor, musicologist and music producer (and earlier as a singer). But as Clive will tell you, although he knows the vague outline, he has "lost all the details."

Clive has been more capable of real and significant conversations in recent years than in the days when he was very scared and angry in the first ten years. He has some awareness of the passage of time as he speaks of his uncle and parents in the past tense (his uncle died in 2003 and after my giving him the news, which upset him as they were close, I don't remember him speaking of Uncle Geoff in the present tense again). Also, if asked to guess how long since his illness, he will guess at least 20 years (in fact 25) and he has always had a rough idea. Again, he has no feeling of knowing, but if asked to guess he is usually spot on.

One other pathological instance that can be attributed to a selective compromise of the autobiographical self is a condition known as anosognosia. Following damage to a region of the *right* cerebral hemisphere that includes the somatosensory cortices and the motor cortices, usually caused by stroke, the patients exhibit a blatant paralysis of the left limb, especially the arm. Yet they repeatedly "forget" that they are paralyzed. No matter how many times they are told that their left arm does not move, when asked, they will still claim, quite sincerely, that it does move. They fail to integrate the information corresponding to the paralysis in the ongoing process of their life history. Their biography is not updated for such facts, even if they do know, for example, that they have suffered a stroke and are admitted to a hospital. This literal oblivion to such blatant realities is responsible for the apparent indifference toward their health condition and for their lack of motivation to participate in the rehabilitation they so need.

I must add that when patients suffer equivalent damage to the *left* cerebral hemisphere, there *never* is anosognosia. In other words, the mechanism by which we update our biographies relative to the aspects of our body having to do with the musculoskeletal system *require* the aggregate of somatosensory cortices located in the *right cerebral hemisphere*.

Seizures arising within this same system can cause a bizarre and fortunately temporary condition: asomatognosia. The patients maintain a sense of self and retain aspects of visceral perception but suddenly and for a brief period are not able to perceive the musculoskeletal aspects of their bodies.

One last comment regarding the pathologies of consciousness. It has recently been suggested that the insular cortices would be the basis for conscious awareness of feeling states and, by extension, of consciousness.[20] It would follow from such a hypothesis that bilateral damage to the insular cortices would cause a devastating disturbance of consciousness. We know from direct observation that this is not true and that patients with bilateral insular damage have normal core self and perfectly active conscious minds.

10

Putting It Together

By Way of Summary

It is time to put together the seemingly disparate facts and hypotheses about brain and consciousness introduced in the previous three chapters. I propose to begin by addressing a number of questions that are likely to have been raised in readers' minds.

1. Granted that consciousness does not reside in a brain center, is it the case that conscious mental states are predominantly based in some brain sectors more than others? My answer is a definite yes. I believe the contents of consciousness that we can access are assembled mostly in the image space of early cortical regions and upper brain stem, the brain's composite "performance space." What happens in that space, however, is continuously engineered by interactions with the dispositional space that spontaneously organizes images as a function of ongoing perception and past memories. At any given moment, the conscious brain works globally, but it does so in an *anatomically differentiated* manner.

2. Any mention of human consciousness conjures up visions of the highly developed cerebral cortex, and yet I have written many pages relating human consciousness to the humble brain

stem. Am I prepared to ignore received wisdom and designate the brain stem as the lead partner in the conscious process? Not quite. Human consciousness requires both the cerebral cortex and the brain stem. The cerebral cortex cannot do it all alone.

3. We have a growing understanding of how neuron circuits work. Mental states have been linked to the firing rates of neurons and to the synchronization of neuron circuits by oscillatory activity. We also know that compared to other species human brains have a larger number and greater specialization of brain areas, especially in the cerebral cortex; that the human cerebral cortex (along with those of apes, whales, and elephants) contains some unusually large neurons known as Von Economo neurons; and that the dendritic branches of some prefrontal cortex neurons in primates are especially abundant compared to those of other cortical regions and of other species. Are these newly discovered features sufficient to explain human consciousness? The answer is no. These features help explain the richness of the human mind, the vast panorama that we can access when minds become conscious as a result of varied self processes. But in themselves they do not explain how self and subjectivity are generated, even if some of these same features play a role in self mechanisms.

4. Feelings are often ignored in accounts of consciousness. Can there be consciousness without feelings? No. Introspectively, human experience always involves feelings. Of course, the merits of introspection can be questioned, but regarding this issue what we need to explain is why conscious states appear to us the way they do, even if the appearance is misleading.

5. I hypothesized that feeling states are generated largely by brain-stem neural systems as a result of their particular design and position vis-à-vis the body. A skeptic may well conclude that I have not answered the question of why feelings feel the way they do, let alone why they feel like anything at all. Here I both agree

and disagree. I have certainly not provided a comprehensive explanation for the making of feelings, but I am advancing a specific hypothesis, aspects of which can be put to the test.

Neither the ideas discussed in this book nor the ideas presented by several colleagues working in this area can be said to solve the mysteries surrounding brain and consciousness. But the current work includes several researchable hypotheses. Only time will tell if they can deliver on their promise.

The Neurology of Consciousness

I see the neurology of consciousness as organized around the brain structures involved in generating the lead triad of wakefulness, mind, and self. Three major anatomical divisions—the brain stem, the thalamus, and the cerebral cortex—are principally involved, but one must caution that there are no direct alignments between each anatomical division and each component of the triad. All three divisions contribute to some aspect of wakefulness, mind, and self.

THE BRAIN STEM

The brain-stem nuclei provide a good illustration of the multitasking required of each division. To be sure, the brain-stem nuclei contribute to wakefulness, in partnership with the hypothalamus, but they are also responsible for constructing the protoself and for generating primordial feelings. Accordingly, significant aspects of the core self are implemented in the brain stem, and once the conscious mind becomes established, the brain stem assists with the governance of attention. In all of these tasks the brain stem cooperates with the thalamus and the cerebral cortex.

To gain a better picture of how the brain stem contributes to the conscious mind, we need to look more closely into the components involved in these operations. An analysis of brain-stem neuroanatomy reveals several sectors of nuclei. The sector located at the bottom of the stem's vertical axis, largely in the medulla oblongata, contains the nuclei that are concerned with basic visceral regulation, notably breathing and cardiac function. Substantial destruction of these nuclei spells death. Above that level, in the pons and in the mesencephalon, we find the nuclei whose damage has been associated with coma and vegetative state rather than death. Roughly, this is the sector that runs vertically from the midlevel of the pons to the top of the mesencephalon; it occupies the back part of the stem rather than the front, behind a vertical line that separates the back half of the brain stem from the front. Two more structures are also part of the brain stem: the tectum and the hypothalamus. The tectum is the ensemble made by the superior and inferior colliculi that we discussed in Chapter 3; architecturally, it provides a sort of roof at the top and at the back of the brain stem. Besides their role in movement related to perception, the colliculi play a role in the coordination and integration of images. The hypothalamus is located immediately above the brain stem, but its deep involvement in life regulation and intricate interactions with brain-stem nuclei justify its inclusion in the brain-stem family. We already addressed the role of the hypothalamus when we dealt with wakefulness in Chapter 8 (please refer to Figure 8.3).

The idea that certain sectors of the stem would be critical for consciousness, but others would not, came from a classical observation made by two distinguished neurologists, Fred Plum and Jerome Posner. They believed that only damage located above the level of the midpons was associated with coma and vegetative state.[1] I turned the idea into a specific hypothesis by proposing a reason for this level setting: when we consider the brain stem from the perspective of brain regions located higher up in the nervous system, we discover that only above the level of the midpons does the collecting of *whole-body information* become complete. At lower levels of the brain stem or spinal cord, the nervous system can avail itself only of *partial* information about the body. This is because the midpons level is the level at which the trigeminal nerve penetrates the brain stem, bringing with it information about the top sector of the body—face and everything behind it, scalp, cranium, and meninges. Only above this level does the brain possess all the information it needs to create comprehensive maps of the whole body and, within such maps, generate the representation of the relatively invariant aspects of the interior that help define the protoself. Below that level the brain has not yet collected all the signals it needs to create a moment-to-moment representation of the entire body.

This hypothesis was tested in a study that Josef Parvizi and I conducted in comatose patients aimed at investigating the location of their brain damage using magnetic resonance. It revealed that coma was associated only with damage above the trigeminal level entry. The study entirely supported Plum and Posner's early observation, which had been based on postmortem material in the age before brain imaging was available.[2]

Early in the history of consciousness research, the association between damage to this region and coma/vegetative state was taken to mean that the resulting dysfunction disrupted wakefulness or vigilance. The cerebral cortex was no longer energized and made active. Deprived of its wakefulness component, the mind was no longer conscious. The identification of a network of locally interactive neurons that projected upward, as a unit, toward the thalamus and cerebral cortex made this simple idea all the more plausible. Even the name given to this system of projections—the ascending reticular activating system, or ARAS—captured the notion successfully.[3] (Again, please refer to Figure 8.3. In Figure 8.3 the ARAS is contained within "other brain-stem nuclei," as noted in the legend.)

The existence of such a system has been thoroughly confirmed, and we know that its projections are aimed at the intralaminar nuclei of the thalamus, which in turn project to the cerebral cortices, including the PMCs. But this is not the whole story. In parallel with classical nuclei such as the cuneiform and pontis oralis, which are where the ARAS originates, there is a rich collection of other nuclei that includes those involved in the management of internal body states: the locus coeruleus, the ventral tegmental nuclei, and the raphe nuclei, respectively responsible for the release of norepinephrine, dopamine, and serotonin in certain sectors of the cerebral cortex and basal forebrain. The projections from these nuclei bypass the thalamus.

Among the nuclei involved in body state management, we find the nucleus tractus solitarius (NTS) and the parabrachial nucleus (PBN), whose significance was discussed in Chapters 3, 4, and 5 relative to the creation of a first line of bodily feelings, the primordial feelings. The upper brain stem also includes the nuclei of the periaqueductal gray (PAG), whose activity results in the behavioral and chemical responses that are part and parcel of life regulation and, as part of that role, execute the emotions. The PAG nuclei are closely interlocked with those of the PBN and the NTS and also with the deep layers of the superior colliculi, which are likely to play a coordinating role in the construction of the core self. This complicated anatomy tells us that while the classical nuclei and the ascending activating systems are no doubt associated with wake and sleep cycles, the remainder of the brain-stem nuclei participate in other equally important functions relevant to consciousness, namely, the housing of the standards for biological value; the representation of the organism's interior on the basis of which the protoself is assembled and primordial feeling states are generated; and the critical first stages in the construction of the core self, which has consequences for the governance of attention.[4]

In brief, reflection on this profusion of functional roles reveals a shared dedication to the management of life. But the idea that the work of these nuclei is confined to the regulation of viscera, metabolism, and wakefulness does not do justice to the results they achieve. They manage life in a far broader way. This is the neural home of biological value, and biological value has a pervasive influence throughout the brain, in terms of structure and operations. In all likelihood, this is the place where the process of making mind begins, in the form of primordial feelings, and it is apparent that the process that makes the conscious mind a reality, the self, also originates here. Even the coordinating efforts of the deep layers of the superior colliculi get into the act and lend a hand.

THE THALAMUS

Consciousness is often described as the result of massive integration of signals in the brain, across many regions; in that description, the role of the thalamus is most prominent. Without a doubt the thalamus contributes importantly to the creation of the background fabric of the mind and to the endgame we call the conscious mind. But can we be more specific about its roles?

Like the brain stem, the thalamus contributes to all components of the conscious mind triad. One set of thalamic nuclei is essential for wakefulness and bridges brain stem to cortex; another brings in the inputs with which cortical maps can be assembled; the remainder assists with the sort of integration without which a complex mind is not conceivable, let alone a mind with a self in it.

I have always resisted venturing into the thalamus, and I am even more cautious today. What little knowledge I have of the huge collection of thalamic nuclei, I owe to the very few experts on this structure.[5] Still, some of the roles played by the thalamus are not in question and can be reviewed here. The thalamus serves as a way station for information that's collected from the body and destined for the cerebral cortex. This includes all the channels that ferry signals about the body and about the world, from pain and temperature to touch, hearing, and vision. All signals bound for the cortex stop at thalamic relay nuclei and change into tracks that take them to their destinations in varied cities of the cerebral cortex. Only smell manages to escape the thalamic attractor and wafts to the cerebral cortex, as it were, via nonthalamic channels.

The thalamus also deals with the signals required to wake up the entire cerebral cortex or put it to sleep—this is done by neuron projections from the reticular formation that I mentioned earlier. Their signals change paths at the intralaminar nuclei, and the PMCs are a major destination.

But no less importantly—and far more specifically when it comes to consciousness—the thalamus serves as a coordinator of cortical activities, a function that depends on the fact that several thalamic nuclei that talk to the cerebral cortex are in turn talked back to and that moment-to-moment recursive loops can be formed. Such thalamic nuclei interconnect parts of the cerebral cortex, distant as well as close. The purpose of the connectivity is not to deliver primary sensory information but instead to *interassociate* information.

In this close interplay between thalamus and cortex, the thalamus is likely to facilitate the simultaneous or sequenced activation of spatially separate neural sites, thus bringing them together in coherent patterns. Such activations are responsible for the flow of images in one's stream of thought, the images that become conscious when they succeed in generating core self pulses. This coordinating role is likely to depend on a back-and-forth between the associative thalamic nuclei and the CDRegions that are, in of themselves, also involved in coordinating cortical activities. The thalamus, in short, both relays critical information to the cerebral cortex and massively interassociates cortical information. The cerebral cortex cannot operate without the thalamus, the two having coevolved and been inseparably joined from early development.

THE CEREBRAL CORTEX

We finally turn to the current pinnacle of neural evolution, the human cerebral cortex. In interplay with the thalamus and brain stem, the cortex keeps us awake and helps select what we attend to. In interplay with the brain stem and thalamus, the cortex constructs the maps that become mind. In interplay with the brain stem and thalamus, the cortex helps generate the core self. Last, using the records of past activity stored in its vast memory banks, the cerebral cortex constructs our biography, replete with the experience of the physical and social environments we have inhabited. The cortex provides us with an identity and places us at the center of the wondrous, forward-moving spectacle that is our conscious mind.[6]

Assembling the consciousness show is such a cooperative effort that it would be unrealistic to single out any particular partner. We cannot engender the autobiographical aspects of self that so define human consciousness without invoking the exuberant growth of convergence-divergence regions that dominate cortical neuroanatomy and neurophysiology. Autobiography could not arise without the seminal contributions of the brain stem toward the protoself, or without the brain stem's obligate consorting with the body proper, or without the brain-wide recursive integration brought in by the thalamus.

But while we need to acknowledge the ensemble work of these major players, it is advisable to resist conceptions that trade the specificity of the contributing parts for an emphasis on functionally indistinct, brain-wide neural operations. In terms of its brain basis, the globalized nature of the conscious mind is undeniable. But we have a chance of finding out more about the relative contributions of brain components to the overall process, thanks to neuroanatomically driven research.

The Anatomical Bottleneck Behind the Conscious Mind

The three main divisions we just outlined and their spatial articulation tell a tale of anatomical disproportions and functional alliances that only an evolutionary perspective can help explain. One does not need to be a neuroanatomist to realize the strange mismatch between the size of the human cerebral cortex and that of the human brain stem.

In essence, adjusted for body size, the basic design of the human brain stem dates back to reptilian times. But the human cerebral cortex is a different story. The cerebral cortex of mammals has expanded enormously, not merely in size but in architectural design, especially in the primate version.

Because of its mastery in the role of life regulator, the brain stem has long been the recipient and local processor of the information needed to represent the body and control its life. And as it discharged this ancient and important role, in species whose cerebral cortex was minimal or absent, the brain stem also developed the machinery required for elementary mind processes and even consciousness, via the protoself and core self mechanisms. The brain stem continues to carry out these same functions in humans today. On the other hand, the greater complexity of the cerebral cortex has enabled detailed image-making, expanded memory capacity, imagination, reasoning, and eventually language. Now comes the big problem: notwithstanding the anatomical and functional expansion of the cerebral cortex, the functions of the brain stem were *not* duplicated in the cortical structures. The consequence of this economic division of roles is a fatal and complete interdependence of brain stem and cortex. They are *forced* to cooperate with each other.

Brain evolution was faced with a major anatomo-functional bottleneck, but natural selection predictably solved it. Given that the brain stem was still being asked to guarantee the full scope of life regulation *and* the foundations of consciousness for the entire nervous system, a way had to be found of ensuring that the brain stem influenced the cerebral cortex *and,* just as important, that the activities of the cerebral cortex influenced the brain stem, most critically, of course, when it came to the construction of the core self. This is all the more important when we think that most external objects exist as images only in the cerebral cortex and cannot be fully imaged in the brain stem.

This is where the thalamus came to the rescue, as the enabler of an accommodation. The thalamus accomplishes a dissemination of signals from the brain stem to a widespread territory of the cortical mantle. In turn, the hugely expanded cerebral cortex, both directly and with the assistance of subcortical nuclei such as those in amygdalae and basal ganglia, funnels signals to the small-scale brain stem. Maybe in the end the thalamus is best described as the marriage broker of the oddest couple.

The brain-stem–cortex mismatch is likely to have imposed limitations on the development of cognitive abilities in general and on our consciousness in particular. Intriguingly, as cognition changes under pressures such as the digital revolution, the mismatch may have a lot to say about the way the human mind evolves. In my formulation the brain stem will remain a provider of the fundamental aspects of consciousness, because it is the first and indispensable provider of primordial feelings. Increased cognitive demands have made the interplay between the cortex and brain stem a bit rough and brutal, or, to put it in kinder words, they have made the access to the wellspring of feeling more difficult. Something may yet have to give.

I said it would be foolish to take sides and favor one of the three divisions in the process of the making of consciousness. And yet one has to agree that the brain-stem component has a functional precedence, that it remains an entirely indispensable part of the puzzle, and that, for that very reason as well as for its modest size and jam-packed anatomy, it is the most vulnerable to pathology among the big three divisions. This much needs to be said, if only because in the wars of consciousness the cerebral cortex tends to get the upper hand.

From the Ensemble Work of Large Anatomical Divisions to the Work of Neurons

Up to this point, I have attempted to explain the emergence of a conscious mind largely from the perspective of components that can be identified with the naked eye, including the small nuclei of the brain stem and thalamus. What the naked eye does not see, however, is the millions of neurons that make up the networks or systems within those structures, nor the numerous small groupings of such neurons that contribute to the overall effort of making a mind with a self. The ensemble work of the large anatomical divisions is built on the ensemble work of components of gradually smaller scale, all the way down to small circuits of neurons. In this downward anatomical trend, there are smaller and smaller regions of the cerebral cortex, along with their retinues of cable work connecting them to other brain sites; there are smaller and smaller nuclei wired in particular ways to other nuclei and to regions of the cortex; last, at the bottom of the hierarchy, we find the small neuron circuits, the microscopic building blocks whose momentary spatial patterns of activity create minds. The conscious mind is built from the brain's nested, hierarchical componentiality.

It is generally assumed that the firing of neurons linked by synapses within microscopic circuits gives rise to the basic phenomena of mind-making, conveniently called the "protophenomena" of cognition. It is also thought that scaling up a large number of such phenomena results in the making of the maps we know as images, and that a part of that scaling-up process depends on the synchronization of the separate protophenomena, as suggested in Chapter 3.

Now, is it enough to combine the microevents of protocognition and synchrony and scale them up across a nested hierarchy distributed within the three neuroanatomical divisions we discussed earlier? In the above account, protocognition from neural microevents is scaled up to the conscious mind, but feeling is omitted. Is there an equivalent "protofeeling" built from neural microevents and scaled up in parallel with protocognition?

In all the proposals advanced in the previous chapters, feeling was presented as an obligate and founding partner for the conscious mind, but nothing was said about its possible microorigins. As proposed earlier, we obtain spontaneous feelings from the protoself, and those feelings give rise, hybridly, to a first flicker of mind and a first flicker of subjectivity. Later, we invoked feelings of knowing to separate self from nonself and to help generate a proper core self. Eventually, we built an autobiographical self from multiple such feeling components. Feelings were presented as the other side of the cognition coin, but their emergence was placed at the systems level. I invoked the unique, resonantly looped and bonded relationship of brain stem to body, and the exhaustive, recursive combination of body signals in the upper brain stem, as sources of qualitatively distinct body feelings. That may well be sufficient to explain how feelings arise. However, it is reasonable to wonder about an additional feature. If we place the origin of images, in general, at the microlevel, with small neuron circuits generating fragments of protocognition, why should we not accord the special class of images we call feelings the same treatment and have them begin within or close to those same small circuits? In the next section, I suggest that feelings may have such a humble origin. Protofeelings would then be scaled up across nested hierarchies into larger circuitry, in this case the circuitry of the upper-brain-stem tegmentum, where additional processing would result in primordial feelings.

When We Feel Our Perceptions

Anyone interested in the matters of brain, mind, and consciousness has heard of qualia and has an opinion regarding what neuroscience can do about the issue: take it seriously and try to deal with it, or consider it intractable and table it, or dismiss it outright. As the reader can see, I take the issue seriously. But first, given that the concept of qualia is somewhat slippery, let us try to make clear what the issue is.[7]

In the text ahead, qualia is treated as a composite of two problems. In one, qualia refers to the feelings that are an obligate part of any subjective experience—some shade of pleasure or its absence, some shade of pain or discomfort, well-being, or lack thereof. I call this the Qualia I problem. The other problem cuts deeper. If subjective experiences are accompanied by feelings, how are feeling states engendered in the first place? This goes beyond the question of how any experience acquires specific sense qualities in our mind, such as the sound of a cello, the taste of wine, or the blueness of the sea. It addresses a blunter question: Why should the construction of perceptual maps, which are physical, neuro-chemical events, feel like something? Why should they feel like anything at all? This is the Qualia II problem.

Qualia I

No set of conscious images of any kind and on any topic ever fails to be accompanied by an obedient choir of emotions and consequent feelings. As I am looking at the Pacific Ocean dressed in its morning suit, protected by a soft, gray sky, I am not just *seeing,* I am also *emoting* to this majestic beauty and feeling a whole array of physiological changes that translate, now that you ask, into a quiet state of well-being. This is happening through no deliberation of mine, and I have no power to prevent the feelings, any more than I had any power to initiate them. They came, they are, and they will stay in some modulation or other, as long as the same conscious object remains in sight and as long as my reflections keep them in some sort of reverberation.

I like to think of Qualia I as music, as a score that accompanies the *remainder* of the ongoing mental process, but noting that the performance is *within* the mental process too. When the main object in my consciousness is not the ocean but an actual music piece, then there are two musical tracks going in my mind, one with the Bach piece that is playing right now and another with the *music-like* track with which I react to the actual music in the language of emotion and feeling. That is none other than Qualia I for a musical performance—call it music on music. Perhaps polyphonic music was inspired by an intuition of this accumulation of parallel "musical" lines in one's mind.

In a small range of real-life situations, the *obligate* Qualia I accompaniment may be reduced or even fail to materialize. The most benign would come from the effect of any drug capable of shutting down emotional responsivity—think of a tranquilizer like Valium, an antidepressant like Prozac, or even a β blocker such as propranolol, all of which, given enough dosage, dampen one's ability to respond emotionally and consequently to experience emotional feelings.

Emotional feelings also fail to materialize in a common pathological situation, depression, in which aspects of positive feeling are notoriously absent and in which even negative feelings such as sadness may be dampened so severely that the result is an affectively blunted state.

How does the brain produce the requisite Qualia I effect? As we saw in Chapter 5, in parallel with the devices of perception that map any object you may wish, and in parallel with the regions, which display such maps, the brain is equipped with a variety of structures that *respond* to signals from those maps by producing emotions, out of which arise subsequent feelings. Examples of such hot-button regions include structures we encountered earlier: the famous amygdala; an almost-as-famous part of the prefrontal cortex known as the ventromedial sector; and an array of nuclei in the basal forebrain and the brain stem.

The way emotions are triggered is intriguing, as we saw earlier. The image-making regions can signal to any of the emotion-triggering regions, directly or after further processing. If the configuration of signals fits the profile that a given region is wired to respond to—that is, if it qualifies as an emotionally competent stimulus—the result is the triggering of a cascade of events, enacted in other parts of the brain and, subsequently, in the body itself, the result of which is an emotion. The perceptual readout of the emotion is a feeling.

The secret behind my composite experience of this moment is the brain's capacity to respond to the *same content* (say, my image of the Pacific Ocean) at *different sites* and in *parallel*. From one brain site I get the emotional process that culminates in a feeling of well-being; from other brain sites I get several ideas about today's weather (the sky does not have quite the typical marine layer; it has more of a cotton fluff appearance, an uneven set of clouds) or about the sea (it can have imposing majesty or welcoming openness depending on the light and the wind, not to mention one's own mood), and so forth.

A normal conscious state usually contains a number of objects to be known, rarely one, and it treats them in a more or less integrated fashion, although hardly ever in the democratic style that would accord equal conscious space and equal time to every object. The fact that different images have different values results in uneven image enhancements. In turn, the uneven enhancement generates an "ordering" of images best described as a spontaneous form of editing. Part of the process of according different values to different images relies on the emotions they provoke and the feelings that ensue in the background of the conscious field—the subtle but not discardable Qualia I response. This is why, although the qualia issue is traditionally regarded as part of the consciousness problem, I believe it belongs more appropriately under the mind rubric. Qualia I responses concern objects being processed in mind and add another element to the mind. I do not regard the Qualia I problem as a mystery.

Qualia II

The Qualia II problem centers on the more perplexing question: why should perceptual maps, which are neural and physical events, feel like anything at all? To attempt a layered answer, begin by focusing on the feeling state that I regard as simultaneous foundation of mind and self, namely, the primordial feelings that describe the state of the organism's interior. I need to start here because of the proposed solution for the Qualia I problem: if feelings regarding the organism's state are the obligate accompaniment of all perceptual maps, then we must first explain the origin of those very feelings.

The front line of the explanation takes into consideration some critical facts. Feeling states first arise from the operation of a few brain-stem nuclei that are highly interconnected among themselves and that are the recipients of highly complex, integrated signals transmitted from the organism's interior. In the process of using body signals to regulate life, the activity of the nuclei transforms those body signals. The transformation is further enhanced by the fact that the signals occur in a looped circuit whereby the body communicates to the central nervous system and the latter responds to the body's messages. The signals are not separable from the organism states where they originate. The ensemble constitutes a dynamic, bonded unit. I hypothesize that this unit enacts a functional fusion of body states and perceptual states, such that the dividing line between the two can no longer be drawn. Neurons in charge of conveying to the brain signals about the body's interior would have such an intimate association with interior structures that the signals conveyed would not be merely *about* the state of the flesh but literally extensions of the flesh. Neurons would imitate life so thoroughly that they would become one with it. In brief, in the complex interconnectivity of these brain-stem nuclei, one would find the beginning of an explanation for why feelings—in this case, primordial feelings—feel like something.

However, as I suggested in the previous section, perhaps we can attempt to go deeper into the small neuron circuit level. The fact that neurons are differentiations of other living cells, both functionally distinct and yet organically similar, gives this idea a foothold. Neurons are not microchips receiving signals from the body. The sensory neurons charged with interoception are body cells of a specialized kind receiving signals from other body cells. Moreover, there are aspects of cell life that suggest the presence of forerunners of a "feeling" function. Unicellular organisms are "sensitive" to threatening intrusions. Poke an amoeba, and it will shrink away from the poke. Poke a paramecium, and it will swim away from the poke. We can observe such behaviors and are comfortable to describe them as "attitudes," knowing full well that the cells do not know what they are doing in the sense that we know what we do when we evade a threat. But what about the other side of this behavior, namely, the cell's internal state? The cell does not have a brain, let alone a mind to "feel" the pokes, and yet it responds because something changed in its interior. Transpose the situation to neurons, and therein could reside the physical state whose modulation and amplification, via larger and larger circuits of cells, could yield a *protofeeling,* the honorable counterpart of the protocognition that arises at the same level.

Neurons do have such response capabilities. Take, for example, their inherent "sensitivity" or "irritability." Rodolfo Llinás has used this clue to propose that feelings arise from the specialized sensory functions of neurons but scaled up to the large number of neurons that are part of a circuit.[8] This is my argument as well, similar to the idea I advanced in Chapter 2 regarding the building of a "collective will to live," as expressed in the self process, from the attitudes of numerous single cells joined cooperatively in an organism. Such an idea draws on the notion of the summing up of cellular contributions: large numbers of muscular cells join forces, literally, by contracting simultaneously and producing a major singular and focused force.

There are intriguing nuances to this idea. The specialization of neurons relative to other body cells comes, in good part, from the fact that neurons, along with muscle cells, are excitable. Excitability is a property that derives from a cell membrane in which local permeability for charged ions is allowed to travel from region to region over the distance of an axon. N. D. Cook suggests that the temporary but repeated opening up of the cell membrane is a violation of the nearly hermetic seal that protects life in the neuron's interior and that such vulnerability would be a good candidate for the creation of a moment of protofeeling.[9]

I am by no means affirming that this is how feelings arise, but I regard this line of inquiry as worth pursuing. Finally, I note that these ideas should *not* be confused with the well-known effort of locating the origins of consciousness at the level of neurons, thanks to quantum effects.[10]

. . .

Another layer of the answer as to why perceptual maps of the body should feel like anything calls for evolutionary reasoning. If perceptual maps of the body are to be effective in leading an organism toward avoidance of pain and seeking of pleasure, they should not only feel like something, they actually *ought* to feel like something. The neural construction of pain and pleasure states must have been arrived at early in evolution and must have played a critical role in its course. It was probably drawn on the body-brain fusion that I have emphasized. Notably, prior to the appearance of nervous systems, unbrained organisms already had well-defined body states that necessarily corresponded to what we came to experience as pain and pleasure. The arrival of nervous systems would have spelled a way of portraying such states with detailed neural signals while holding neural and bodily aspects tightly bonded to each other.

A related aspect of the answer points to the functional divide between pleasure and pain states, which are correlated, respectively, with optimal and smooth life-managing operations, in the case of pleasure, and impeded, problem-ridden life-managing operations, in the case of pain. Those extreme ends of the range are associated with the release of particular chemical molecules that have an effect on the body proper (on metabolism, on muscular contraction) and on the brain (where they can modulate the processing of newly assembled as well as recalled perceptual maps). Other reasons aside, pleasure and pain should feel different because they are mappings of very different body states, just as a certain red is different from a particular blue because they have different wavelengths and the voices of sopranos are different from those of baritones because their sound frequencies are higher.

It is often overlooked that information from the body's interior is conveyed directly to the brain by numerous chemical molecules that course in the bloodstream and bathe parts of the brain that are devoid of blood-brain barrier, namely, the area postrema in the brain stem and a variety of regions known collectively as the circumventricular organs. To call the number of potentially active molecules "numerous" is not an exaggeration since the basic list includes dozens of examples (the usual transmitter/modulator suspects—the inevitable norepinephrine, dopamine, serotonin, acetycholine—as well as a wide range of hormones such as steroids and insulin, and opioids). As the blood bathes these receptive areas, the suitable molecules directly activate neurons. This is how, for example, a toxic molecule acting on the area postrema can lead to a practical reaction such as vomiting. But what else do the signals that arise in such areas end up causing? A reasonable guess is that they cause or modulate feelings. Projections from these regions are highly concentrated on the nucleus tractus solitarius but reach out widely to other nuclei in the brain stem, hypothalamus, and thalamus and to the cerebral cortex.

Beyond the issue of feelings, the remainder of the Qualia II problem seems more approachable. Take visual maps, for example. Visual maps are sketches of visual properties, shape, color, movement, depth. Interconnecting such maps—cross-fertilizing their signals, as it were—is the right prescription for producing a blended, multidimensional visual scene. If one takes this blend and adds to it information from the visual portal—to the effect that the flesh around the eyes is involved in the process—and a component of feeling, it is reasonable to expect a full-blown, properly "qualied" experience of what is being seen.

What can we add to this complexity such that the qualities of a percept are indeed distinctive? One thing has to do with the sensory portals involved in gathering the information. Changes in the sensory portals play a role in the buildup of perspective, as we saw, but they also contribute to the construction of perceptual quality. How? We know the distinctive sound of Yo-Yo Ma's playing, and we know where the sound maps are created in the brain, but we hear the sounds both *in* our ears and *with* our ears. In all probability, we feel sounds in our ears because our brains are assiduously mapping *both* the information that comes to the sensory probe—from the entire auditory signaling chain including the cochlea—and the slew of co-occurring signals coming from the apparatus that surrounds the sensory device. In the case of hearing, this includes the epithelium (skin) covering our ears and the external ear canal, along with the tympanic membrane and the tissues holding the system of ossicles that transmit mechanical vibrations to the cochlea. To this we must add the small and not-so-small head and neck movements that we constantly make in an automatic effort to adjust the body to the sound sources. This is the auditory equivalent of the notable changes that occur in the eyeball and the surrounding muscles and skin when we are in the process of looking and seeing, and it adds qualitative texture to the percepts.

The feel of smelling or tasting or touching arises via the same sort of mechanism. For example, our nasal mucosa contains olfactory nerve endings that respond quite directly to the conformation of chemical molecules in odorants—that is how we come to map scents and how we deliver jasmine or Chanel N° 19 for their encounter with our self. But *where* we feel the smell arises from other nerve endings in the nasal mucosa, those that are irritated when you put too much wasabi on the sushi and are forced to sneeze.

Finally, we note that there are back projections from the brain aimed at the body's periphery, including the periphery that contains specialized sensory devices. This could well accomplish for a sensory process such as hearing a milder version of what the brain-stem–body loop accomplishes for feeling: a functional linkage that bridges the gulf between the brain and the starting point of the sensory chains in the body's end-organ periphery. Such a loop might enable another reverberating process. The input cascades aimed at the brain would be complemented by output cascades aimed at the very "flesh" where the signals originated, thus contributing to the integration of inner and outer worlds. We know that such arrangements exist, the auditory system being a prime example. The cochlea receives feedback signals from within the brain, so much so that when the feedback mechanism is unbalanced, the cochlea's hair cells can actually *emit* tones rather than convey them, as they are normally supposed to. We need to know more about the circuitry of the sensory devices.[11]

I believe the foregoing accounts for a substantial part of the problem as it succeeds in bringing together in the mind three kinds of maps: (1) maps of a particular sense generated by the appropriate sensory device, that is, sight, sound, smell, and so forth; (2) maps of the activity in the sensory portal within which the sensory device is embedded in the body; and (3) maps of the emotional-feeling reactions to the maps generated under (1) and (2), that is, Qualia I responses. Those percepts would come to be as they are when different kinds of sensory signals are brought together in mind-making maps of the brain stem or cerebral cortex.[12]

Qualia and Self

How do Qualia I and Qualia II fit in the process of self? Since both aspects of qualia round up the construction of the mind, qualia is part of the contents that come to be known as the self process, the self construction illuminating the mind construction. But somewhat paradoxically, Qualia II is also the grounding for the protoself and thus sits astride mind and self, in a hybrid transition. The neural design that enables qualia provides the brain with *felt* perceptions, a sense of pure experience. After a protagonist is added to the process, the experience is claimed by its newly minted owner, the self.

Unfinished Business

The business of understanding how the brain makes a conscious mind remains unfinished. The mystery of consciousness is still a mystery, although it is being pushed back a bit. But it is too soon to declare defeat.

Discussions of the neurology of consciousness and of the mind-brain problem usually suffer from two blatant underestimations. One consists of not giving proper due to the wealth of detail and organization of the body proper, the facts that the body is replete with micro-nooks and microcrannies and that microworlds of form and function can be signaled to the brain, mapped, and the result put to work for a variety of purposes. The most likely first purpose of those signals is regulatory—the brain needs to receive information describing the state of body systems so that it can organize, nonconsciously or consciously, an appropriate response. Feelings of emotion are the obvious result of such signaling, although feelings have come to loom large in our conscious life and social relationships. In the same way, it is quite possible, indeed likely, that other body processes, some already known, others to be discovered, will turn out to influence our conscious experiences at many levels.

The other underestimation pertains to the brain itself. The idea that we have a firm grasp of what the brain is and what it does is pure folly, but we always know more than we did the year before and much, much more than one decade ago. Problems that seem intolerably mysterious and unbearably hard are likely to be amenable to biological account, the question being not if but when.

PART IV

Long After
Consciousness

I I

Living with Consciousness

Why Consciousness Prevailed

Traits and functions rise or fall in the history of life depending on how much they contribute to the success of living organisms. The most direct way of explaining why consciousness has prevailed in evolution is to say that it has contributed significantly to the survival of the species so equipped. Consciousness came, saw, and conquered. It has flourished. It seems to be here to stay.

What did consciousness actually contribute? The answer is a large variety of apparent and not-so-apparent advantages in the management of life. Even at the simplest levels, consciousness helps the optimization of responses to environmental conditions. As processed in the conscious mind, images provide details about the environment, and those details can be used to increase the precision of a much-needed response, for example, the exact movement that will neutralize a threat or guarantee the capture of a prey. But image precision is only a part of the advantage of a conscious mind. The lion's share of the advantage, I suspect, comes from the fact that in a conscious mind the processing of environmental images is *oriented* by a particular set of internal images, those of the subject's living organism as represented in the self. The self focuses the mind process, it imbues the adventure of encountering other objects and events with a motivation, it infuses the exploration of the world outside the brain with a *concern* for the first and foremost problem facing the organism: the successful regulation of life. That concern is naturally generated by the self process, whose foundation lies in bodily feelings, primordial and modified. The spontaneously, intrinsically feeling self signals directly, as a result of the valence and intensity of its affective states, the degree of concern and need that are present at every moment.

As the process of consciousness became more complex, and as co-evolved functions of memory, reasoning, and language were brought into play, further benefits of consciousness were introduced. Those benefits relate largely to planning and deliberation. The advantages here are legion. It became possible to survey the possible future and to either delay or inhibit automatic responses. An example of this evolutionarily novel capacity is delayed gratification, the calculated trading of something good now for something better later—or the forgoing of something good now when the survey of the future suggests that it will cause something bad as well. This is the trend of consciousness that brought us a finer management of basic homeostasis and, ultimately, the beginnings of sociocultural homeostasis (to which I will turn later in this chapter).

Plenty of conscious, highly successful behaviors are present in many nonhuman species with complex enough brains: the examples are evident all around us, most spectacularly in mammals. In humans, however, thanks to expanded memory, reasoning, and language, consciousness has reached its current peak. I suggest that the peak came from the strengthening of the knower self and of its ability to reveal the predicaments and opportunities of the human condition. Some may say that in that revelation lies a tragic loss, of innocence no less, for all that the revelation tells us of the flaws of nature and of the drama we face, for all the temptations it lays down before human eyes, for all the evil it unmasks. Be that as it may, it is not for us to choose. Consciousness certainly has allowed the growth of knowledge and the development of science and technology, two ways in which we can attempt to manage the predicaments and opportunities laid bare by the human conscious state.

Self and the Issue of Control

Any discussion of the advantages of consciousness must consider mounting evidence to the effect that, on many occasions, the execution of our actions is controlled by nonconscious processes. This does happen frequently enough, in all sorts of settings, and it deserves attention. It is apparent in the execution of skills, from driving a car to playing a musical instrument, and it is constantly present in our social interactions.

The evidence for nonconscious participation in our actions, solid and not so solid, can be easily misinterpreted. It is easy to downplay the value of self-directed conscious control, when it has been shown in numerous experiments, beginning with those of Benjamin Libet and including those of Dan Wegner and Patrick Haggard, that one's subjective impression of when or what initiated an action can be proven wrong.[1] It is just as easy to use such facts, along with evidence from social psychology, as an argument for the need to revise the traditional notion of human responsibility. If factors unknown to our conscious reasoning influence the shape of our acts, are we really responsible for our actions?

But the situation is far less problematic than it may seem from such superficial and unjustified reactions to findings whose interpretation is still being discussed. First, the reality of nonconscious processing and the fact that it can exert control over one's behavior are not in question. Not only that, such nonconscious control is a welcome reality from which we draw palpable advantages, as we shall see. Second, nonconscious processes are, in substantial part and in varied ways, under *conscious* guidance. In other words, there are two kinds of control of actions, conscious and nonconscious, but the nonconscious control can be partly shaped by the conscious variety. Human childhood and adolescence take the inordinate amount of time that they do because it takes a long, long time to educate the nonconscious processes of our brain and to create, within that nonconscious brain space, a form of control that can, more or less faithfully, operate according to conscious intentions and goals. We can describe this slow education as a process of transferring part of the conscious control to an unconscious server, not as a surrender of conscious control to the unconscious forces that, to be sure, can wreak havoc in human behavior. Patricia Churchland has argued this position convincingly.[2]

Consciousness is not devalued by the presence of nonconscious processes. Instead, the reach of consciousness is amplified. And, assuming the presence of a normally functioning brain, the degree of one's responsibility for an action is not necessarily diminished by the presence of healthy and robust nonconscious execution of some actions.

In the end, the relationship between conscious and nonconscious processes is one more example of the odd functional partnerships that emerge as a result of coevolving processes. Of necessity, consciousness and direct conscious control of actions emerged after nonconscious minds were in place, running the show with plenty of good results but not always. The show could be improved. Consciousness came of age by first restraining part of the nonconscious executives and then exploring them mercilessly to carry out preplanned, predecided actions. Nonconscious processes became a suitable and convenient means to execute behavior and give consciousness more time for further analysis and planning.

When we walk home thinking about the solution of a problem rather than about the route we take, but still do get home safe and sound, we have accepted the benefits of a nonconscious skill that was acquired in many previous conscious exercises, following a learning curve. While we were walking home, all that our consciousness needed to monitor was the general goal of the trip. The rest of our conscious processes were free for creative use.

Much the same applies to the professional behaviors of musicians and athletes. Their conscious processing is focused on achieving goals, reaching certain marks at certain epochs, avoiding some perils of execution, and detecting unforeseen circumstances. The rest is practice, practice, practice, the second nature that can guide you to Carnegie Hall.

Last, the conscious-unconscious cooperative interplay also applies in full to moral behaviors. Moral behaviors are a skill set, acquired over repeated practice sessions and over a long time, informed by consciously articulated principles and reasons but otherwise "second-natured" into the cognitive unconscious.

In conclusion, what is meant by conscious deliberation has little to do with the ability to control actions in the moment and everything to do with the ability to plan ahead and decide which actions we want or do not want to carry out. Conscious deliberation is largely about decisions taken over extended periods of time, as much as days or weeks in the case of some decisions, and rarely less than minutes or seconds. It is not about split-second decisions. Common knowledge regards lightning-speed choices as "thoughtless" and "automatic."[3] Conscious deliberation is about *reflection over knowledge*. We apply reflection and knowledge when we decide on important matters in our lives. We use conscious deliberation to govern our loves and friendships, our education, our professional activities, our relations to others. Decisions pertaining to moral behavior, narrowly or broadly defined, involve conscious deliberation and take place over extended time periods. Not only that, such decisions are processed in an offline mental space that overwhelms external perception. The subject at the center of conscious deliberations, the self in charge of the prospection of the future, is often distracted from external perception, inattentive to its vagaries. And there is a very good reason for this distraction in terms of brain physiology: the image-processing brain space, as we have seen, is the sum total of early sensory cortices; this same space needs to be shared by conscious reflection processes *and* direct perception; it is hardly up to the task of handling both without favoring one or the other.

Conscious deliberation, under the guidance of a robust self built on an organized autobiography and a defined identity, is a major consequence of consciousness, precisely the kind of achievement that gives the lie to the notion that consciousness is a useless epiphenomenon, a decoration without which brains would run the life-management business just as effectively and without the hassle. We cannot run our kind of life, in the physical and social environments that have become the human habitat, without reflective, conscious deliberation. But it is also the case that the products of conscious deliberation are significantly limited by a large array of nonconscious biases, some biologically set, some culturally acquired, and that the nonconscious control of action is also an issue to contend with.

Still, most important decisions are taken long before the time of execution, within the conscious mind, when they can be simulated and tested and where conscious control can potentially minimize the effect of nonconscious biases. Eventually the exercise of decisions can be honed into a skill with the help of nonconscious mind processing, the submerged operations of our mind in matters of general knowledge and reasoning often referred to as the cognitive unconscious. Conscious decisions begin with reflection, simulation, and testing in the conscious mind; that process can be completed and rehearsed in the nonconscious mind, from which freshly selected actions can be executed. The conscious as well as the nonconscious components of this complex and fragile decision and execution device can be derailed by the machinery of appetites and desires, in which case a last recourse veto is not likely to be effective. Split-second vetoes remind us of a well-known recommendation on the matter of drug addiction: "Just say no." This strategy may be adequate when one has to preempt an innocuous finger movement, but not when one needs to stop an action urged by a strong desire or appetite, precisely the kind posed by any addiction to drugs, alcohol, attractive foods, or sex. Successful nay-saying requires a lengthy conscious preparation.

An Aside on the Unconscious

Thanks to the fact that our brains have successfully combined the new governance made possible by consciousness with the old governance that consisted of unconscious, automatic regulation, nonconscious brain processes are up to the tasks they are supposed to perform on behalf of conscious decisions. Some suitable evidence can be gleaned from a remarkable study by the Dutch psychologist Ap Dijksterhuis.[4] To appreciate the importance of the results, we need to describe the setting. Dijksterhuis asked the normal subjects of his experiment to make purchasing decisions in two different conditions. In one condition, they applied mostly conscious deliberation; in the other, as a result of manipulated distraction, they could not deliberate consciously.

There were two kinds of items to purchase. One consisted of trivial household items, such as toasters and hand towels; the other consisted of big-ticket items, such as cars or houses. For either kind, the subject was given ample information about the pros and cons of each item, a sort of consumer report complete with price tag. Such information would have come in handy when asked to pick the "best" possible item for purchase. But when decision time came, Dijksterhuis allowed some subjects to study the item information for three minutes before making a choice, while he denied that privilege to the others and distracted them during those same three minutes. For both kinds of items, trite and nontrite, subjects were tested in both conditions, with an attentive three-minute study or with a distraction.

What would you predict regarding the quality of decisions? A perfectly reasonable prediction would be that when it comes to the trivial household items, subjects would make good picks in either conscious or unconscious deliberation, given the low import and complexity of the problem. Deciding between two toasters, even if you are fussy, is hardly rocket science. But, regarding the big-ticket items—like which four-door sedan to buy—one would expect that the subjects allowed to study the information would make the more successful decisions.

The results were surprisingly different from these predictions. Decisions made without conscious predeliberation fared better for both kinds of item but especially for the big-ticket items. The superficial conclusion is as follows: if you are buying a car or a house, get acquainted with the facts, but then don't fret and worry about minute comparisons along the matrix of possible advantages or disadvantages. Just do it. So much for the glories of conscious deliberation.

Needless to say, the intriguing results should not discourage anyone from conscious deliberation. What they do suggest is that nonconscious processes are capable of some sort of reasoning, far more than they are usually thought to be, and that this reasoning, once it has been properly trained by past experience and when time is scarce, may lead to beneficial decisions. In the circumstances of the experiment, the attentive, conscious pondering that goes on, especially with the big-ticket items, does not yield the best result. The high number of variables under consideration and the restricted space of conscious reasoning—restricted by the limited number of items that can be attended to at any given time—reduce the probability of making the best choice given the limited time window. The unconscious space, on the contrary, has a far larger capacity. It can hold and manipulate many variables, potentially producing the best choice in a small window of time.

Besides what it tells us about nonconscious processing in general, the Dijksterhuis study points to other important issues. One regards the amount of time needed for a decision. Perhaps you could pick the absolute best restaurant for tonight if you had all afternoon to examine the latest food reviews, the cost of items on the menus, and the locations, and compare these to your preferences, your mood, and the state of your bank account. But you do not have the entire afternoon. Time counts, and you must apportion only a "reasonable" amount of time to the decision. Reasonableness depends, of course, on the importance of the matter you are deciding. Given that you do not have all the time in the world and rather than making a huge investment in massive computation, a few shortcuts are in order. And the good news is that past emotional records will help you with the shortcuts and that our cognitive unconscious is a good provider of such records.

All this goes to say that I very much like the notion that our cognitive unconscious is capable of reasoning and has a larger "space" for operations than the conscious counterpart. But a critical element for the explanation of these results relates to the subject's prior emotional experience with items similar to the varied big-ticket items in the experiment. The nonconscious space is wide open and suitable for this covert manipulation, but it works to one's advantage largely because certain options are nonconsciously marked by a bias connected to previously learned emotional-feeling factors. I believe that the conclusions on the merits of *un*-consciousness are correct, but our notion of what goes on beneath the glassy surface of consciousness is much enriched when we factor emotion and feeling into the nonconscious processes.

The Dijksterhuis experiment illustrates the combination of unconscious and conscious powers. Unconscious processing alone could not do the job. In these experiments, unconscious processes do a lot of work, but the subjects have benefited from years of conscious deliberation during which their nonconscious processes have been repeatedly trained. Moreover, while nonconscious processes do their due diligence, the subjects remain fully conscious. Unconscious patients under anesthesia or in coma do not make decisions about the real world any more than they enjoy sex. Again, it is the felicitous synergy of the covert and overt levels that carries the day. We feed on the cognitive unconscious quite regularly, throughout the day, and discreetly outsource a number of jobs, including the execution of responses, to its expertise.

Outsourcing expertise to the nonconscious space is what we do when we hone a skill so finely that we are no longer aware of the technical steps needed to be skillful. We develop skills in the clear light of consciousness, but then we let them go underground, into the roomy basement of our minds, where they do not clutter the exiguous square footage of conscious reflection space.

The Dijksterhuis experiment adds a flourish to an ongoing research effort regarding the role of nonconscious influences in decision tasks. Early in that effort, our research group had presented decisive evidence in this regard.[5] For example, we showed that when normal subjects are playing a card game that involves both gains and losses under conditions of risk and uncertainty, the players begin to adopt a winning strategy slightly ahead of being able to articulate why they are doing so. For some minutes preceding their adoption of the advantageous strategy, the subjects' brains produce differential psychophysiological responses whenever they ponder taking a card from one of the bad decks, those that promote losses, while the prospect of lifting a card from a good deck generates no such response. The beauty of the result resides with the fact that the psychophysiological responses, which, in the original study were measured with skin conductance, are *not* perceivable by either the subject or the naked eyes of an observer. They occur under the radar of the subject's consciousness, just as stealthily as the behavioral drift toward the winning strategy.[6]

What exactly is going on is not entirely clear yet, but whatever it is, in-the-moment consciousness is not a requirement. It may be that the nonconscious equivalent of a conscious gut feeling "jolts" the decision-making process, as it were, biasing the nonconscious computation and preventing the selection of the wrong item. In all likelihood, there is an important reasoning process going on nonconsciously, in the subterranean mind, and the reasoning produces results without the intervening steps ever being known. Whatever the process is, it produces the equivalent of an *intuition* without the "aha" acknowledgment that the solution has arrived, just a quiet delivery of the solution.

The evidence for nonconscious processing has increased unabated. Our economic decisions are not guided by pure rationality and are significantly influenced by powerful biases such as the aversion to losses and the delight in gains.[7] The way we interact with others is influenced by a large array of biases having to do with gender, race, manners, accents, and attire. The setting of the interaction brings its own set of biases, linked to familiarity and design. The concerns and emotions we were experiencing prior to the interaction play an important role too, as does the hour of the day: Are we hungry? Are we sated? We express or give indirect signs of preferences for human faces at lightning speed without having had time to process consciously the data that would have backed up a corresponding reasoned inference, which is all the more reason to be extra careful with important decisions in our personal and civic lives.[8] To let the unconscious sway of past emotion guide your choice of a house is fine, provided you stop and reflect carefully on what the unconscious is offering you as an option before you sign the contract. You may conclude that the choice is not valid based on the reanalysis of the data, regardless of how you intuitively judged the situation, because, for example, your past experiences in this domain are atypical, biased, or insufficient. This is all the more important if you are voting in an election or on a jury. One of the major problems faced by voters in political elections and in courtroom trials is the strength of emotional/nonconscious factors. The power of nonconscious, emotional factors is so well recognized that a perfectly monstrous machinery of electoral influence has developed as an industry over the past few decades, along with less publicized but equally sophisticated methods of influential jury selection.

Reflection and reassessment, fact checks, and reconsideration are of the essence. Here is a great occasion to invest in extra decision time, preferably before entering the voting booth or handing your vote to the jury foreman.

All these findings exemplify situations in which nonconscious influences, emotional or not, and nonconscious reasoning steps have a bearing on the outcome of a task. But the subjects are very much conscious when they are given the premises of the task, as well as when the decision occurs, and they are informed of the outcomes of their actions. It is clear that these are examples of nonconscious components of otherwise conscious decisions. They let us glean the complexity and variety of mechanisms behind the facade of allegedly perfect conscious control, but they do not deny our deliberative powers and do not free us from responsibility for our actions.

A Note on the Genomic Unconscious

A brief note is in order regarding the genomic unconscious, one of the hidden forces that conscious deliberation needs to contend with. What do I mean by genomic unconscious? Quite simply, the colossal number of instructions that are contained in our genome and that guide the construction of the organism with the distinctive features of our phenotype, in both body proper and brain, and that further assist with the operation of the organism. The basic design of our brain circuitries is instructed by the genome, and that basic design contains the very first repertoire of nonconscious know-how with which our organisms can be governed. The know-how has to do first and foremost with life regulation, issues of life and death, and reproduction; but precisely because of the centrality of those issues, the design promotes a number of behaviors that may appear to be decided by conscious cognition but are in fact driven by nonconscious dispositions. The spontaneous preferences one manifests early in life, regarding food and drink and mates and habitats, are driven in part from the genomic unconscious, although they can be modulated and modified by individual experience throughout development.

Psychology has long recognized the existence of unconscious foundations of behavior and studied them under the rubrics of instinct, automatic behaviors, drives, and motivations. What has changed recently is the realization that the early placement of such dispositions in the human brain is under considerable genetic influence and that, notwithstanding all the shaping and remodeling we undertake as conscious individuals, the thematic scope of such dispositions is wide and their pervasiveness astonishing. This is especially notable regarding some of the dispositions on which cultural structures have been built. The genetic unconscious had something to say about the early shaping of the arts, from music and painting to poetry. It had something to do with the early structuring of the social space, including its conventions and rules. It had something to do, as both Freud and Jung certainly sensed, with many aspects of human sexuality. It had a lot to contribute to the fundamental narratives of religion and to the time-honored plots of plays and novels, which revolve in no small part around the force of genomically inspired emotional programs. Blindly set jealousy, impervious to common sense, hard evidence, and reason, drives Othello to kill the perfectly innocent Desdemona, and Karenin to punish the adulterous Anna Karenina so harshly. Iago's monumental malevolence would probably not have worked were it not for Othello's natural vulnerability to jealousy. The cognitive asymmetry of sexuality in men and women, many parameters of which are engraved in our genomes, lurks behind the behavior of these characters and keeps them ever modern. The intense male aggression of Achilles, Hector, and Ulysses has equally deep roots in the genetic unconscious. The same may be said of two characters, Oedipus and Hamlet, destroyed either by the breaking of the incest taboo or by the unstated inclination to break it. The Freudian interpretation of these timeless characters merges with their evolutionary origins, pointing to some highly frequent features of human nature. Theater and the novel, as well as film, their twentieth-century heir, have greatly benefited from the genomic unconscious.

The genomic unconscious is partly responsible for the sameness that hallmarks the repertoire of human behavior. How remarkable it is, then, that we consistently break away from monotonous universals and instead, by dint of artistry or the sheer magic of a human encounter, create an infinite set of life variations that delights and astonishes.

The Feeling of Conscious Will

How frequently are we guided by a well-rehearsed cognitive unconscious, trained under the supervision of conscious reflection to observe consciously conceived ideals, wants, and plans? How frequently are we guided by unconscious, deeply set, biologically ancient biases, appetites, desires? I suspect that most of us, weak but well-meaning sinners, operate on both registers, more on one or more on the other, depending on the situation and the hour of the day.

Whatever register we operate on, somewhat virtuous or somewhat not, in-the-moment acting is inevitably accompanied by the impression, sometimes false, sometimes not, that we acted there and then, under full conscious control our self plunged headlong into whatever we did. That impression is a *feeling,* a feeling that arises when our organisms engage in a new perception or initiate a new action, none other than the feeling of knowing that I discussed earlier as part and parcel of the assembled self. Someone who shares this view is Dan Wegner, who describes conscious will as "the somatic marker of personal authorship, an emotion that authenticates the action's owner as the self. With the feeling of doing an act, we get a conscious sensation of will attached to the action."[9] In other words, we are not mere "conscious automata," as T. H. Huxley considered us to be, a century ago, unable to control our existence.[10] When the mind is informed of the actions taken by our organism, the feeling associated with the information signifies that the actions were engendered by our self. Both information and authentication of ongoing actions are essential to motivate the deliberation of future actions. Without that sort of felt, validated information, we would not be able to assume moral responsibility for the actions taken by our organism.

Educating the Cognitive Unconscious

Greater control over the vagaries of human behavior can come only from an accumulation of knowledge and from consideration of the discovered facts. Taking the time to analyze facts, to evaluate the outcome of decisions, and to ponder the emotional results of those decisions is the path to building a practical guide otherwise known as wisdom. On the basis of wisdom, we can deliberate and hope to steer our behavior within the frame of cultural conventions and ethical rules that have informed our biographies and the world we live in. We can also react to those conventions and rules, face the conflict that ensues when we disagree with them, and even attempt to modify them. A good example is the conflict faced by conscientious objectors.

No less important, we need to be aware of the peculiar hurdle faced by our consciously deliberated decisions—they have to find a way into the cognitive unconscious in order to permeate the action machinery— and we need to facilitate that influence. One way to transpose the hurdle would be the intense conscious rehearsal of the procedures and actions we wish to see nonconsciously realized, a process of repeated practice that results in mastering a *performing skill,* a consciously composed psychological action program gone underground.

I am not inventing anything new here but merely outlining a practical mechanism deduced from what I presume the neural operations of decision and action must be like. For millennia wise leaders have turned to a comparable solution when they asked followers to observe disciplined rituals whose side effect must have been a gradual imposition of consciously willed decisions on nonconscious action processes. Not surprisingly, those rituals often involved the creation of heightened emotions, even pain, an empirically discovered means of etching the desired mechanism in the human mind. What I am envisioning, however, extends well beyond religious and civic rituals to encompass day-to-day life matters bearing on a variety of areas. I am thinking, in particular, about matters of health and social behavior. Our insufficient education of nonconscious processes probably explains, for example, why so many of us fail miserably to do what we are supposed to do regarding diet and exercise. We *think* we are in control, but we often are not, and the epidemics of obesity, hypertension, and heart disease prove that we are not. Our biological makeup inclines us to consume what we should not, but so do the cultural traditions that have drawn on that biological makeup and been shaped by it, and even the advertising industry that exploits it. There is no conspiracy here. It is only natural. Perhaps this is a good place for ritualized skill building, if that is what it takes.

The same applies to the epidemic of drug addiction. One reason so many individuals become addicted to all manner of drugs, not to mention alcohol, has to do with the pressures of homeostasis. In the natural course of a day, we inevitably face frustrations, anxieties, and difficulties that throw homeostasis off balance and consequently may make us feel unwell, perhaps anguished, discouraged, or sad. One effect of the so-called substances of abuse is to restore the lost balance rapidly and transiently. How do they do so? I believe they change the felt image that the brain is currently forming of its body. The state of off-balance homeostasis is neurally represented as an impeded, troubled body landscape. After certain drugs, at certain dosages, the brain represents a more smoothly functioning organism. The suffering that corresponds to the former felt image morphs into temporary pleasure. The brain's appetite system has been hijacked, and the eventual result is not quite the sought-after rebalancing of homeostasis, at least not for long. Nonetheless, rejecting the possibility of rapid correction of suffering takes enormous effort, even for those who already know that the correction is short-lived and the consequences of the choice may be dire. In the framework I have outlined, there is an obvious reason for this state of affairs. The nonconscious homeostatic demand is in natural control and can be opposed only by a well-trained and powerful counterforce. Spinoza seems to have had the right idea when he said that an emotion with negative consequences could be countered only by another, more powerful emotion. What this possibly means is that merely training the nonconscious process to politely decline is hardly a solution. The nonconscious device must be trained by the conscious mind to deliver an emotional counterpunch.

Brain and Justice

Biologically informed conceptions of conscious and unconscious control are relevant to how we live and especially to how we should live.

But perhaps the relevance is nowhere more important than on issues that regard social behavior—in particular, the sector of social behavior known as moral behavior—and the breaking of the social agreements codified in laws.

Civilization, and in particular the aspect of civilization that has to do with justice, revolves around the notion that humans are conscious in ways in which animals are not. By and large, cultures have evolved justice systems that take a commonsense approach to the complexities of decision-making and aim at protecting societies from those who violate established laws. Understandably, and with rare exceptions, the weight given to evidence coming from brain science and cognitive science has been negligible.

Now there is a growing fear that evidence regarding brain function, as it becomes more widely known, may undermine the application of laws, something that legal systems have by and large avoided by not taking such evidence into account. But the response has to be nuanced. The fact that everyone capable of knowing is responsible for his actions does not mean that the neurobiology of consciousness is irrelevant to the process of justice and to the process of education charged with preparing future adults to an adaptive social existence. On the contrary, lawyers, judges, legislators, policy-makers, and educators need to acquaint themselves with the neurobiology of consciousness and decision-making. This is important to promote the writing of realistic laws and to prepare future generations for responsible control of their actions.

In certain cases of brain dysfunction, even the best-exercised deliberation may fail to overpower forces either nonconscious or conscious, it does not matter. We are barely beginning to glean the profile of such cases, but we do know, for example, that patients with certain kinds of prefrontal damage may be unable to control their impulsivity. The way in which such individuals control their actions is not normal. How are they to be judged when they come under the purview of justice? As criminals or as neurological patients? Perhaps both, I would say. Their neurological disease should in no way pardon their actions, even if it may explain aspects of a crime. But if they are neurologically sick, then they are indeed patients, and society needs to handle them accordingly. A current tragedy in this regard is that we are just beginning to understand these facets of neurological disease; once the conditions are diagnosed, we have very little to offer in terms of treatment. But that in no way limits society's responsibility regarding the understanding and public debate of the available knowledge, and the need for further research on these matters.[11]

Some other patients, in whom the prefrontal damage is concentrated on the ventromedial sector, judge hypothetical moral dilemmas in a very practical, utilitarian manner that has little or no use for the better angels of the human spirit. And when such patients are confronted with, say, a hypothetical case of attempted murder that did not result in death in spite of a murderous intent, they do not judge the situation as significantly different from that of an accidental and unintended killing. In fact, they may even find the former situation more permissible.[12] The way in which such individuals understand motives, intents, and consequences is unconventional, to say the least, even if in their daily lives they probably would not harm a fly. We still have much to learn about how the human brain processes judgments of behavior and controls actions.

Nature and Culture

The history of life is shaped like a tree with numerous branches, each leading out to different species. Even species that are not at the end of a high branch can be superbly intelligent within their own zoological neighborhood. Their achievements should be judged relative to that neighborhood. Still, when we take the long view of the tree of life, we cannot fail to recognize that organisms do progress from simple to complex. In that perspective it is reasonable to wonder when consciousness

appeared in the history of life. What did it do for life? If we scan biological evolution as an unpremeditated march up the tree of life, the
sensible answer is that consciousness appeared quite late, high in the
tree. There is no sign of consciousness in the primordial soup or in bacteria, in unicellular or simple multicellular organisms, in fungi or plants,
all interesting organisms that exhibit elaborate life-regulation devices,
precisely those devices whose accomplishments consciousness will
improve upon at a later date. None of those organisms has a brain, let
alone a mind. In the absence of neurons, behavior is limited and mind
not possible, and if there is no mind, there is no consciousness as such,
only precursors of consciousness.

When neurons make their appearance, life changes remarkably.
Neurons emerge as a variation on the theme of other body cells. They
are made up of the same components as other cells, they go about their
general business in the very same way, and yet they are special. Neurons
become carriers of signals, processing devices capable of transmitting
messages and receiving them. By virtue of those signaling capabilities,
neurons organize themselves in complex circuits and networks. In turn,
circuits and networks represent events occurring in other cells and,
directly or indirectly, influence the function of other cells and even
their own function. Neurons are through and through *about* other cells
in the body, although they do not lose their body cell status just because
they have acquired the ability to transmit signals electrochemically, dispatch those signals to a variety of places in an organism, and constitute
circuits and systems of enormous complexity. They are body cells,
exquisitely dependent on nutrients as all body cells are, differing mostly
in their ability to play tricks that other body cells cannot play, and
firmly set on their attitude to live long, if possible as long as their owners. The body-brain separation has been somewhat exaggerated since
the neurons that make up the brain *are* body cells, something that does
have a bearing on the body-mind problem.

Once neurons are in place inside organisms capable of movement,
life changes in a way that nature denied to plants. A relentless progres-

sion of functional complexity begins, from ever more elaborate behaviors to mind processes and eventually to consciousness. One secret behind this complexification is now clear. It has to do both with the sheer number of neurons available in a given organism and, just as important, with the patterns in which they are organized as circuits of gradually larger and larger scales, all the way up into macroscopic brain regions that form systems with intricate functional articulations. The combined significance of neuron numbers and organization pattern is the reason why it is not possible to approach the problems of behavior and mind by relying exclusively on the investigation of individual neurons, or of the molecules that act on them, or of the genes involved in the running of their life. Studying individual neurons, microcircuits, molecules, and genes is indispensable in order to understand the problem comprehensively. But the mind and behavior of apes and humans are so different because of the *number* of brain elements and the *pattern* of organization of those elements.

Nervous systems developed as managers of life and curators of biological value, assisted at first by unbrained dispositions but eventually by images, that is, minds. The emergence of mind produced spectacular improvements in life regulation for numerous species, even when images lacked fine detail and lasted only during the perceptual moment, entirely vanishing thereafter. The brains of social insects are an example of those achievements, amazingly sophisticated and yet somewhat inflexible, vulnerable to interruptions of their behavioral sequences, and not yet capable of holding representations in a temporary working memory space. Minded behavior became very complex in numerous nonhuman species, but it is arguable that the flexibility and creativity that hallmark human performance could not have emerged from a generic mind alone. The mind had to be protagonized, had to be enriched by a self process arising in its midst.

Once self comes to mind, the game of life changes, albeit timidly at first. Images of the internal and external worlds can be organized in a cohesive way around the protoself and become oriented by the homeostatic requirements of the organism. Then the devices of reward and punishment and drives and motivations, which had been shaping the life process in earlier stages of evolution, help with the development of complex emotions. Then social intelligence begins to be flexible. The eventual presence of the core self is followed by an expansion of mental processing space, of conventional memory and recall, of working memory, and of reasoning. Life regulation focuses on a gradually more well-defined individual. Eventually the autobiographical self emerges, and with its arrival the regulation of life changes radically.

If nature can be regarded as indifferent, careless, and unconscionable, then human consciousness creates the possibility of questioning nature's ways. The emergence of human consciousness is associated with evolutionary developments in brain, behavior, and mind that ultimately lead to the creation of culture, a radical novelty in the sweep of natural history. The appearance of neurons, with its attending diversification of behavior and paving of the way into minds, constitutes a momentous event in the grand trajectory. But the appearance of conscious brains eventually capable of flexible self-reflection is the next momentous event. It is the opening of the way into a rebellious, albeit imperfect response to the dictates of a careless nature.

How did the independent and rebellious mind develop? One can only speculate, and the pages ahead are a mere sketch of an immensely complex picture that cannot be accommodated in a single book, let alone a chapter. Nonetheless we can be certain that the rebel did not develop suddenly. Minds constituted by maps of diverse sensory modalities were helpful in improving life regulation, but even when the maps became properly felt mental images, they were not independent, let alone rebellious. Felt images of the organism's interior made for improved survival and created a potentially nice spectacle but there was no one to watch it. When minds first added a core self to their stock, which is when consciousness really began, we were getting closer to the mark but not quite yet there. A simple protagonist was a clear advantage, because it generated a firm connection between life-regulation needs and the profusion of mental images that the brain was forming about the world around it. The guidance of behavior was optimized. But the independence I am talking about could surface only once the self was complex enough to reveal a fuller picture of the human condition, once living organisms could learn that pain and loss were at stake but so were pleasure and flourishing and folly, once there were questions to be asked about the human past and the human future, once imagination could show how possibly to reduce suffering, minimize loss, and increase the probability of happiness and fancy. That is when the rebel began to take human existence in new directions, some defiant, some accommodating, but all based on thinking through knowledge, mythical knowledge at first, scientific knowledge later, but knowledge nonetheless.

Self Comes to Mind

How wonderful it would be to discover where and when the robust self came to mind and began generating the biological revolution called culture. But in spite of the ongoing research efforts of those who interpret and date the human records that have survived time, we are not able to answer such questions. It is certain that the self matured slowly and gradually but unevenly, and that the process was taking place in several parts of the world, not necessarily at the same time. Still, it is known that our most direct human ancestors were walking the earth about 200,000 years ago, and that around 30,000 years ago humans were producing cave paintings, sculptures, rock carvings, metal castings, and jewelry, and possibly making music. The proposed date for the Chauvet cave, in Ardèche, is 32,000 years ago, and by 17,000 years ago the Lascaux cave was already a Sistine Chapel of sorts, with hundreds of complex paintings and thousands of carvings, in a complex mixture of figures and abstract signs. A mind capable of symbolic processing was obviously at work there. The exact relation between the emergence of language and the explosion of artistic expression and sophisticated tool-making that distinguishes *Homo sapiens* is not known. But we do know that for tens of thousands of years humans had engaged in burials elaborate enough to require special treatment of the dead and the equivalent of tombstones. It is difficult to imagine how such behaviors would have occurred in the absence of an explicit concern for life, a first stab at interpreting life and assigning it value, emotional of course, but intellectual as well. And it is inconceivable that concern or interpretation could arise in the absence of a robust self.

The development of writing, about five thousand years ago, provides a handful of solid evidence, and by the time of the Homeric poems, which are likely to be less than three thousand years old, autobiographical selves had undoubtedly come to human minds. Still, I sympathize with Julian Jaynes's claim that something of great import may have happened to the human mind during the relatively brief interval of time between the events narrated in the *Iliad* and those that make up the *Odyssey*.[13] As knowledge accumulated about humans and about the universe, continued reflection could well have altered the structure of the autobiographical self and led to a closer stitching together of relatively disparate aspects of mind processing; coordination of brain activity, driven first by value and then by reason, was working to our advantage. Be that as it may, the self that I envision as capable of rebelliousness is a recent development, on the order of thousands of years, a mere instant in evolutionary time. That self draws on features of the human brain acquired, in all likelihood, during the long period of the Pleistocene. It depends on the brain's capacity to hold expansive memory records not only of motor skills but also of facts and events, in particular, personal facts and events, those that make up the scaffolding of biography and personhood and individual identity. It depends on the ability to reconstruct and manipulate memory records in a working brain space parallel to the perceptual space, an offline holding area where time can be suspended during a delay and decisions freed from the tyranny of immediate responses. It depends on the brain's ability to produce not only mental representations that imitate reality slavishly and mimetically but also representations that symbolize actions and objects and individuals. The rebellious self depends on the brain's ability to communicate mental states, especially feeling states, through gestures of body and hands, as well as through the voice, in the form of musical tones and verbal language. Last, it depends on the invention of external memory systems parallel to those held by each brain, by which I mean the pictorial representations offered by early painting, carvings, and sculpture, tools, jewelry, funerary architecture, and, long after the emergence of language, written records, certainly the most important variety of external memory until quite recently.

Once autobiographical selves can operate on the basis of knowledge etched in brain circuits and in external records of stone, clay, or paper, humans become capable of hitching their individual biological needs to the accumulated sapience. Thus begins a long process of inquiry, reflection, and response, expressed throughout recorded human history in myths, religions, the arts, and various structures invented to govern social behavior—constructed morality, justice systems, economics, politics, science, and technology. The ultimate consequences of consciousness come by way of memory. This is memory acquired through a filter of biological value and animated by reason.

The Consequences of a Reflective Self

Imagine early humans sometime after verbal language established itself as a means of communication. Imagine conscious individuals whose brains were armed with many of the abilities we find in humans today and who sought much of what we seek today—food, sex, shelter, security, comfort, dignity, perhaps transcendence. In that environment competition for resources was a dominant problem, conflict would be abundant, and cooperation was essential. Reward, punishment, and learning oriented their behaviors. Let us assume that they possessed a range of emotions resembling ours. Attachment, disgust, fear, joy, sadness, and anger were no doubt present, along with emotions that governed sociality such as trust, shame, guilt, compassion, contempt, pride, awe, and admiration. And let us assume these early humans were already animated by an intense curiosity regarding both their physical environment and other living beings, of the same species or not. If twentieth-century studies of relatively isolated tribes are any guide, they were also curious about themselves and told stories about their origin and their destiny. The engines behind such curiosity are relatively easy to envision. Early humans would experience affection and attachment for others with whom they bonded, especially mates and offspring, and they would have experienced the grief that comes from breaking those bonds, or from witnessing others in suffering, or experiencing their own suffering. They also would have experienced and witnessed moments of joy and satisfaction, and times of success at the endeavors of hunting, courtship, securing shelter, war, the raising of the young.

This systematic discovery of the drama of human existence and of its possible compensations was arguably possible only after full-fledged human consciousness developed—a mind with an autobiographical self capable of guiding reflective deliberation and gathering knowledge. Eventually, given the probable intellectual capability of early humans, it is likely that they would have wondered about their status in the universe, something akin to the *where from* and *where to* questions that still haunt us today, thousands of years later. That is when the rebellious self comes of age. That is when myths are developed to account for the human condition and its workings; when social conventions and rules are elaborated, leading to the beginnings of a true morality that sits over and above promoral behaviors such as kin altruism and reciprocal altruism, behaviors that nature had long been exhibiting prior to the emergence of reflective selves; when religious narratives are created from and around myths, aimed both at explaining the reasons behind the drama and at enforcing the new laws designed to reduce it. In brief, reflective consciousness not only improved the revelation of existence but allowed conscious individuals to begin interpreting the condition and taking action.

I suggest that the engine behind these cultural developments is the *homeostatic impulse*. Explanations that rely only on the significant cognitive expansions that bigger and smarter brains produced are insufficient to account for extraordinary developments of culture. In one form or another, the cultural developments manifest the same goal as the form of automated homeostasis to which I have alluded throughout this book. They respond to a detection of imbalance in the life process, and they seek to correct it within the constraints of human biology and of the physical and social environment. The elaboration of moral rules and laws and the development of justice systems responded to the detection of imbalances caused by social behaviors that endangered individuals and the group. The cultural devices created in response to the imbalance aimed at restoring the equilibrium of individuals and of the group. The contribution of economic and political systems, as well as, for example, the development of medicine, responded to functional problems that occurred in the social space and that required correction within that space, lest they compromise the life regulation of the individuals that constituted the group. The imbalances that I am referring to are defined by social and cultural parameters, and the detection of imbalance thus occurs at the high level of the conscious mind, in the brain's stratosphere, rather than at the subcortical level. I call this overall process "sociocultural homeostasis." Neurally speaking, sociocultural homeostasis begins at the cortical level, although the emotional reactions to the imbalance immediately engage basic homeostasis as well, testifying once again to the hybrid life regulation of the human brain, high, then low, then high, in an oscillatory course that frequently flirts with chaos but barely avoids it. Conscious reflection and planning of action introduce new possibilities in the governance of life over and above automated homeostasis, in a remarkable novelty of physiology. Conscious reflection can even question and modulate automated homeostasis and decide on an optimal range of homeostasis at a level higher than needed for survival and more consistently conducive to well-being. The imagined, dreamed-of, anticipated well-being has become an active motivator of human action. Sociocultural homeostasis was added on as a new functional layer of life management, but biological homeostasis remained.

Armed with conscious reflection, organisms whose evolutionary design was centered around life regulation and the tendency toward homeostatic balance invented forms of consolation for those in suffering, rewards for those who helped the sufferers, injunctions for those who caused harm, norms of behavior aimed at preventing harm and promoting good, and a mixture of punishments and preventions, of penalties and praise. The problem of how to make all this wisdom understandable, transmissible, persuasive, enforceable—in a word, of how to make it stick—was faced and a solution found. Storytelling was the solution—storytelling is something brains do, naturally and implicitly. Implicit storytelling has created our selves, and it should be no surprise that it pervades the entire fabric of human societies and cultures. It also should be no surprise that the sociocultural narratives borrowed their authority from mythical beings presumed to have more power and more knowledge than humans, beings whose existence explained all manner of predicaments and whose activity had the ability to offer succor and modify the future. Over the skies of the Fertile Crescent or in storybook Valhalla, those beings have exerted a fascinating hold on the human mind.

Individuals and groups whose brains made them capable of inventing or using such narratives to improve themselves and the societies they lived in became successful enough for the architectural traits of those brains to be selected, individually and groupwise, and for their frequency to increase over generations.[14]

. . .

The idea that there are two broad classes of homeostasis, basic and sociocultural, should not be taken to mean that the latter is a purely "cultural" construction, while the former is "biological." Biology and culture are thoroughly interactive. Sociocultural homeostasis is shaped by the workings of many minds whose brains have first been constructed in a certain way under the guidance of specific genomes. Intriguingly, there is growing evidence that cultural developments can lead to profound modifications in the human genome. For example, the invention of dairy farming and the availability of milk in the diet has led to changes in the genes that permit lactose tolerance.[15]

I suspect that precisely the same homeostatic impulse that shaped the development of myths and religions was behind the emergence of the arts, aided by the same intellectual curiosity and explanatory drive. This may sound ironic given that Freud regarded the arts as an antidote to the neuroses caused by religions, but I mean no irony. The same conditions could indeed give rise to these two developments. If the need to manage life was one of the reasons music, dance, painting, and sculpting first emerged, then the ability to improve communication and organize social life were two other strong reasons and gave the arts additional staying power.

Close your eyes for a moment, and imagine humans of long ago, perhaps even before language made its appearance but mindful and conscious, already equipped with emotions and feelings, already aware of what it is to be sad or to be joyful, to be in danger or to be in safety and comfort, to enjoy gain or suffer loss, to have pleasure or pain. And now imagine how they would have expressed those states of which they were mindful. Perhaps they would intone calls of danger or calls of greeting, calls of gathering, calls of joy, calls of mourning. Perhaps they would hum or even sing, since the human vocal system is an inbuilt musical instrument. Or, for that matter, imagine drumming, given that the chest cavity is a natural drum. Imagine drumming as a mind-concentrating device or as a social organizing tool—a drum to order, a drum to arms—or imagine blowing on a primitive bone flute as a means of magic enchantment, seduction, consolation, playful merriment. It is not Mozart yet, and it is not *Tristan and Isolde,* but a way had been found. Dream some more.

At the birth of arts such as music, dance, and painting, people probably intended to communicate to others information about threats and opportunities, about their own sadness or joy, and about shaping social behavior. But in parallel to communication, the arts would also have produced a homeostatic compensation. If they had not, would they have prevailed? All of this even before the marvelous discovery that when humans were able to produce words and string them together in sentences, not all sounds sounded alike. The sounds had natural accents, and the accents could have relationships in time. Accents could create rhythms, and certain rhythms would create pleasure. Poetry could begin, and the technique could eventually feed back into the practice of music and dance.

The arts could emerge only once brains acquired certain mental features that in all likelihood became established over a long evolutionary period, again the Pleistocene. There are many examples of such features. They include the emotive reaction of pleasure to certain shapes and certain pigments, present in natural objects but applicable to human-made objects as well as to body decoration; the pleasurable reaction to certain features of sounds and to certain kinds of organization of sounds in relation to timbres, to pitches and their relationships, as well as to rhythms. Similarly for the emotive reaction to certain kinds of spatial organization and to landscapes that include open vistas and proximity to water and vegetation.[16]

Art may have begun as a homeostatic device for artist and recipient and as a means of communication. Eventually on the side of the artist and on the side of the audience, the uses became quite varied. Art became a privileged means to transact factual and emotional information deemed to be important for individuals and society, something established in early epic poems, theater, and sculpture. Art also became a means to induce nourishing emotions and feelings, something at which music has excelled through the ages. No less important, art became a way to explore one's own mind and the minds of others, a means to rehearse specific aspects of life, and a means to exercise moral judgment and moral action. Ultimately, because the arts have deep roots in biology and the human body but can elevate humans to the greatest heights of thought and feeling, they became a way into the homeostatic refinement that humans eventually idealized and longed to achieve, the biological counterpart of a spiritual dimension in human affairs.

In brief, the arts prevailed in evolution because they had survival value and contributed to the development of the notion of well-being. They helped cement social groups and promote social organization; they assisted with communication; they compensated for emotional imbalances caused by fear, anger, desire, and grief; and they probably inaugurated the long process of establishing external records of cultural life, as suggested by Chauvet and Lascaux.

It has been suggested that art survived because it made artists more successful at attracting mates; we need only to think of Picasso and smile in agreement. But the arts would have probably prevailed on the basis of their therapeutic value alone.

The arts were an inadequate compensation for human suffering, for unattained happiness, for lost innocence, but they were and are compensation nonetheless, an offset to natural calamities and to the evil that men do. They are one of the remarkable gifts of consciousness to humans.

And what is the ultimate gift of consciousness to humanity? Perhaps the ability to navigate the future in the seas of our imagination, guiding the self craft into a safe and productive harbor. This greatest of all gifts depends, once again, on the intersection of the self and memory. Memory, tempered by personal feeling, is what allows humans to imagine both individual well-being and the compounded well-being of a whole society, and to invent the ways and means of achieving and magnifying that well-being. Memory is responsible for ceaselessly placing the self in an evanescent here and now, between a thoroughly lived past and an anticipated future, perpetually buffeted between the spent yesterdays and the tomorrows that are nothing but possibilities. The future pulls us forward, from a distant vanishing point, and gives us the will to continue the voyage in the *present*. This may be what T. S. Eliot meant when he wrote: "Time past and time future / What might have been and what has been / Point to one end, which is always present."[17]

APPENDIX

When you look at three-dimensional views of the human brain, there is an obvious architectural arrangement that you grasp with the naked eye. The overall pattern is similar from brain to brain, and certain components show up in every brain in the same position. Their relationship is like that of the components of our faces—eyes, mouth, nose. Their exact shape and size are somewhat different in each individual, but the range of variation is limited. There are no human faces in which the eyes are square or in which an eye is larger than the nose or mouth, and symmetry is by and large respected. Comparable restrictions apply to the relative positions of the elements. Like our faces, our brains are extremely similar in terms of the grammatical rules according to which the parts are arranged in space. And yet brains are quite individual. Each brain is unique.

Another aspect of the architecture that is relevant to the ideas in this book, however, is invisible to the naked eye. Lying beneath the surface, it consists of a massive cable work made up of *axons*—the fibers that interconnect neurons. The brain has billions of neurons (about 10^{11}), and those neurons make trillions of connections among themselves (about 10^{15}). Nonetheless, the connections are made according to *patterns,* and not every neuron connects to every other neuron. On the contrary, their meshwork is highly selective. Seen from afar it constitutes a wiring diagram, or many wiring diagrams, depending on the sector of the brain.

Understanding the wiring diagrams is one road to understanding what the brain does and how. But it is not easy because the wiring diagrams undergo considerable changes during development and beyond. We are born with certain connection patterns, put into place under the instruction of our genes. These connections were already influenced by several environmental factors in the womb. After birth individual experiences in unique environments get to work on that first connection pattern, pruning it away, making certain connections strong and others weak, thickening or thinning the cables in the network, under the influence of our own activities. Learning and creating memory are simply the process of chiseling, modeling, shaping, doing, and redoing our individual brain wiring diagrams. The process that began at birth continues until death makes us part with life, or some time before, if Alzheimer's disease disrupts the process.

How does one uncover the design of the wiring diagrams? Until quite recently, research on this problem required brain specimens, largely postmortem material from either humans or experimental animals. Samples of brain tissue would be fixed and stained with identifiable dyes, and very thin slices of tissue could be analyzed under the microscope. There is a venerable tradition of such studies in experimental neuroanatomy, and they have yielded most of the knowledge we have today about the brain's networking. But our knowledge of neuroanatomy remains embarrassingly incomplete, so there is an urgent need for such studies to continue, making use of considerable progress in the available stains and in the power of modern microscopes.

Recently, new possibilities have opened with the use of magnetic resonance methods in living humans. Noninvasive methods such as diffusion imaging are allowing us a first glimpse of in vivo human connection networks. Although the techniques are still far from satisfactory, they promise to yield fascinating revelations.

How do the billions of neurons inside a human brain and the trillions of synapses they form manage to produce not just the actions that constitute behaviors but also minds—minds of which each owner can be conscious and minds that can give rise to cultures? To say that so many neurons and synapses do the job by massive interactivity and by the ensuing complexity is not a good answer. Interactivity and complexity must surely be present, but interactivity and complexity are not amorphous. They derive from the varied designs of local circuit arrangements and the even more varied ways in which such circuits create regions and regions become affiliated in systems. How each region is made, internally, determines its function. A region's location in the overall architecture is important too, because its place in the global plan determines its partners in the system—the regions that talk to a particular region and to which it talks back. To make matters even more complicated, the opposite is also true: to a certain extent the partners that it interacts with determine where its place is going to be. But before we go any further, we should give a brief account of the materials used to construct brain architecture.

BRICKS AND MORTAR

The mind-making brain is made of neural tissue, and neural tissue, like any other living tissue, is made of cells. The principal type of brain cell is the *neuron,* and for reasons that I alluded to in Chapters 1, 2, and 3, the neuron is a distinctive cell in the universe of biology. Neurons and their axons are embedded—*suspended* might be a better term—in a scaffolding made up of another type of brain cell, the *glial cell.* Besides providing neurons with physical support, glial cells also provide part of their nourishment. Neurons cannot survive without glial cells, but everything indicates that neurons are the critical brain unit as far as behavior and mind are concerned.

When neurons use their axons and send messages to muscular fibers, they can produce movements; and when neurons are active within very complex networks of map-making regions, the result is images, the main currency of mental activity. Glial cells, as far as we know, do nothing of the sort, although their full contribution to the operation of

neurons has not been fully elucidated. On a somber note, glial cells are
the origin of the most deadly brain tumors, the gliomas, for which there
is no cure to date. Even worse, for reasons that are entirely unclear, the
incidence of malignant gliomas is rising worldwide, unlike practically
all other malignancies. The other common origin of brain tumors is the
cells of the meninges—the skinlike membranes that cover brain tissue.
Meningiomas tend to be benign, although, by dint of their location and
unchecked growth, they can compromise brain function seriously and
are anything but innocent.

Each neuron has three main anatomical elements: (1) the *cell body,* which
is the cell's powerhouse and includes the cell nucleus and organelles such
as mitochondria (the neuron's genome, its complement of governing
genes, is located within the nucleus, although DNA is also to be found
within mitochondria); (2) the main output fiber, known as the *axon,*
which arises from the cell body; and (3) input fibers, known as *dendrites*
that stick out from the cell body a bit like antlers. Neurons are connected to one another via a border area called the *synapse.* In most
synapses the axon of one neuron makes chemical contact with the dendrites of another.

Neurons can be active (firing) or inactive (not firing), "on" or "off."
The firing consists of producing an electrochemical signal that crosses
the border to another neuron, at the synapse, and makes that other neuron fire too, provided the signal meets the requirements of the other
neuron to fire. The electrochemical signal travels from the neuron's
body down the axon. The synaptic border is located between the end of
an axon and the beginning of another neuron, generally at the dendrite.
There are several minor variations and exceptions to this standard
description, and different kinds of neurons vary in shape and size; but
this outline is acceptable as far as the big picture goes. Each neuron
is so small that one needs the major amplification of a microscope to
see it, and in order to see a synapse one needs an even more powerful

microscope. Still, smallness is relative, entirely in the amplified eye of the beholder. Compared to the molecules that make them up, neurons are truly gigantic creatures.

When neurons "fire," the electric current known as the action potential is propagated away from the cell body and down the axon. The process is very fast—it takes only a handful of milliseconds, which should give an idea of the remarkably different time scales of brain and mind processes. We need *hundreds* of milliseconds to become conscious of a pattern presented to our eyes. We experience feelings in a time scale of *seconds,* that is *thousands* of milliseconds, and *minutes*.

When the firing current arrives at a synapse, it triggers the release of chemicals known as neurotransmitters (glutamate is an example) in the space between two cells, the synaptic cleft. In an excitatory neuron, the cooperative interaction of many other neurons whose synapses are adjacent and that release (or do not) their own transmitter signals, determines whether the next neuron will fire, that is, whether it will produce its own action potential, which will lead to its own neurotransmitter release, and so forth.

Synapses can be strong or weak. Synaptic strength determines whether and how easily impulses will continue to travel into the next neuron. In an excitatory neuron, a strong synapse facilitates impulse travel, while a weak synapse impedes or blocks it.

One critical aspect of learning is the strengthening of a synapse. Strength is translated into ease of firing and thus ease of activation of the neurons downstream. Memory depends on this operation. Our understanding of the neural basis of memory at neuron level can be traced to the seminal ideas of Donald Hebb, who, in the mid–twentieth century, first raised the possibility that learning depended on the strengthening of synapses and the facilitation of the firing of subsequent neurons. He did so on a purely theoretical basis, but his hypothesis was subsequently proven correct. In the past few decades the understanding of learning has deepened to the level of molecular mechanisms and gene expression.

On average each neuron talks to relatively few others, not to most, and never to all. In fact, many neurons talk only to neurons that are close by, within relatively local circuits; others, even if their axons travel for several centimeters, make contact with only a small number of other neurons. Still, depending on where the neuron sits in the overall architecture, it may have more or fewer partners.

The billions of neurons are organized in circuits. Some are very small microcircuits, truly local operations invisible to the naked eye. When many microcircuits are placed together, however, they form a region, with a certain architecture.

The elementary regional architectures come in two varieties: the *nucleus* variety and the *cerebral cortex patch* variety. In a patch of cerebral cortex, the neurons are displayed on two-dimensional surface sheaths stacked in layers. Many of these layers have a fine topographical organization. This is ideal for detailed mapping. In a nucleus of neurons (not to be confused with the cell nucleus inside each neuron), the neurons are usually displayed like grapes inside a bowl, but there are partial exceptions to this rule. The geniculate nuclei and the collicular nuclei, for example, have two-dimensional, curvy layers. Several nuclei have topographical organization as well, which suggests that they can generate coarse maps.

Nuclei contain "know-how." Their circuitry embodies knowledge about how to act or what to do when certain messages make the nucleus active. Because of this dispositional know-how, nucleus activity is indispensable for the management of life in species with smaller brains, those with little or no cerebral cortex and limited map-making abilities. But nuclei are also indispensable for managing life in brains such as ours, where they are responsible for basic management—metabolism, visceral responses, emotions, sexual activity, feelings, *and* aspects of consciousness. The governance of endocrine and immune systems depends on nuclei, and so does affective life. But in humans, a good part of the operation of nuclei is under the influence of the mind, and that means largely, though not entirely, the influence of the cerebral cortex.

Importantly, the separate regions defined by nuclei and by cerebral cortex patches are interconnected. They form, in turn, larger- and larger-scale circuits. Numerous patches of cerebral cortex come to be wired together interactively, but each patch is also wired to subcortical nuclei. Sometimes the patch of cortex is a recipient of signals from a nucleus, or sometimes it is a sender of signals; sometimes it is both recipient and sender. The interactions are especially significant in relation to the myriad nuclei of the thalamus (regarding which the connections to the cerebral cortex tend to be two-way) and in relation to the basal ganglia (regarding which the connections tend to be either downward from cortex or up toward it, but not both).

In sum, neuron circuits constitute cortical regions, if they are arranged in sheaths placed in parallel layers like those of a cake or constitute nuclei, if they are grouped in nonlayered arrangements (but note the exceptions mentioned earlier). Both cortical regions and nuclei are interconnected by axon "projections" to form *systems* and, at gradually higher levels of complexity, *systems of systems*. When bunches of axon projections are large enough to be seen by the naked eye, they are called "pathways." In terms of scale, all neurons and local circuits are microscopic, while all cortical regions, most nuclei, and all systems of systems are macroscopic.

If neurons are the bricks, what is the brain's equivalent of mortar? Quite simply, it is the large number of *glial* cells that I introduced as the scaffolding for the neurons everywhere in the brain. The myelin sheaths that wrap around fast-conducting axons are also glial. They provide protection and insulation for those axons, conforming yet again to the role of mortar. Glial cells are very different from neurons in that they do not have axons and dendrites and do not transmit signals over long distances. In other words, glial cells are not about the other cells in an organism, and their role is neither to regulate nor to represent other cells. The imitative role of neurons does not apply to glial cells. But the roles that glial cells play go beyond mere shelving for neurons. Glial cells intervene in the nutrition of neurons by holding and delivering energy products, for example, and, as suggested earlier, their influence may actually go deeper.

MORE ON THE LARGE-SCALE ARCHITECTURE

The nervous system has central and peripheral divisions. The main component of the *central nervous system* is the *cerebrum,* which is made up of two *cerebral hemispheres,* left and right, joined by the *corpus callosum.* A facetious tale says that the corpus callosum was invented by nature to keep the cerebral hemispheres from sagging. But we know that this thick collection of nerve fibers connects the left and right halves, in both directions, and performs an important integrative role.

The cerebral hemispheres are covered by the cerebral cortex, which is organized in lobes (*occipital, parietal, temporal, and frontal*) and includes a region known as the *cingulate cortex,* visible only on the internal (mesial) surface. Two regions of the cerebral cortex that are not visible at all when one inspects the surface of the cerebellum are the *insular cortex,* buried underneath the frontal and parietal regions; and the *hippocampus,* a special cortical structure hidden in the temporal lobe.

Underneath the cerebral cortex, the central nervous system also includes deep conglomerates of nuclei such as the *basal ganglia,* the *basal forebrain,* the *amygdala,* and the *diencephalon* (a combination of the *thalamus* and the *hypothalamus*). The cerebrum is joined to the spinal cord by the *brain stem,* behind which the *cerebellum* is located with its two hemispheres. Although the hypothalamus is usually mentioned together with the thalamus to constitute the diencephalon, in reality the hypothalamus is functionally closer to the brain stem, with which it shares the most critical aspects of life regulation.

The central nervous system is connected to every point of the body by bundles of axons originating in neurons. (The bundles are known as nerves.) The sum total of all nerves connecting the central nervous system with the periphery and vice versa constitutes the *peripheral nervous system.* Nerves transmit impulses from brain to body and from body to

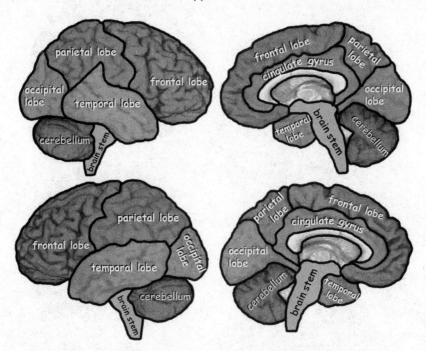

Figure A.1: The large-scale architecture of the human brain shown in a three-dimensional reconstruction of magnetic resonance data. The lateral (external) views of both right and left cerebral hemispheres are shown on the left panels; the medial (internal) views are shown on the right. The white curved structure in the right panels corresponds to the corpus callosum.

brain. One of the oldest and most important sectors of the peripheral nervous system is the *autonomic nervous system,* so called because its operation is largely outside our volitional control. The components of the autonomic nervous system include the *sympathetic, parasympathetic,* and *enteric* systems. The system plays a critical role in life regulation and in emotions and feelings. The brain and the body are also interconnected by chemical molecules such as hormones, which travel in the bloodstream. The ones that travel from brain to body originate in nuclei such as those in the hypothalamus. But chemical molecules also travel in the opposite direction and influence neurons directly at locations such as the area postrema, where the protective blood-brain barrier is missing.

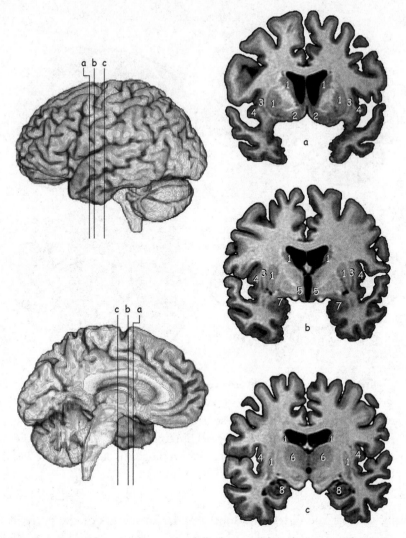

Figure A.2: The panels on the left depict three-dimensional reconstructions of the human brain seen from lateral and medial perspectives (top and bottom, respectively).

The panels on the right depict three sections of the brain volume. The sections were obtained along the lines marked *a*, *b*, and *c*. The sections reveal a number of important brain structures located under the surface: 1 = basal ganglia; 2 = basal forebrain; 3 = claustrum; 4 = insular cortex; 5 = hypothalamus; 6 = thalamus; 7 = amygdala; 8 = hippocampus. The cerebral cortex covers the entire surface of the cerebral hemispheres, including the depth of every sulcus. In the sections, the cerebral cortex appears as a dark rim easily distinguishable from the lighter white matter underneath. The black areas at the center of the sections correspond to the lateral ventricles.

(The blood-brain barrier is a protective shield against certain molecules circulating in the bloodstream.) The area postrema is located in the brain stem, very close to important life-regulating structures such as the parabrachial and periaqueductal nuclei.

When one slices the central nervous system in any direction and looks at the cross-section, one notices a difference between dark and pale sectors. The dark sectors are known as the *gray matter* (although they are more brown than gray), and the pale sectors are known as the *white matter* (which is more tan than white). The gray matter gets its darker hue from the tight packing of many neuron cell bodies; the white matter gets its lighter appearance from the insulating sheaths of the axons that emanate from the cell bodies located in the gray matter. As noted, the insulation is made of myelin and speeds up the conduction of electric current in the axons. Myelin insulation and fast conduction of signals are hallmarks of evolutionarily modern axons. Unmyelinated fibers are quite slow and of older vintage.

The gray matter comes in two varieties. By and large the layered variety is found in the *cerebral cortex,* which envelops the cerebral hemispheres, and in the *cerebellar cortex,* which envelops the cerebellum. The nonlayered variety is made of *nuclei,* the main examples of which were listed earlier: the *basal ganglia* (located in the depth of each cerebral hemisphere and made up of three large nuclei, the caudate, the putamen, and the pallidum); the *amygdala,* a single and sizable lump located in the depth of each temporal lobe; and several aggregations of smaller nuclei that form the *thalamus,* the *hypothalamus,* and the gray sectors of the *brain stem*.

The cerebral cortex is the cerebrum's mantle, covering the surfaces of each cerebral hemisphere, including those that are located in the depth of fissures and sulci, the crevices that give the brain its unique folded appearance. The thickness of the cortex is about three millimeters, and the layers are parallel to one another and to the brain's surface. The evolutionarily modern part of the cerebral cortex is the *neocortex*. The main divisions of the cerebral cortex are designated as lobes:

frontal, temporal, parietal, and occipital. All other gray structures (the various nuclei mentioned earlier and the cerebellum) are subcortical.

In the text I often refer to *early sensory cortices* or to *association cortices* or even to *higher-order association cortices*. The designation *early* has no time connotation at all; it refers to the position occupied by a region in space, along a sensory processing chain. Early sensory cortices are those located near and around the point of entry of peripheral sensory pathways into the cerebral cortex—for example, the point of entry for vision or hearing or touch signals. The early regions tend to be organized concentrically. They play a critical role in producing detailed maps using the signals brought in by the sensory pathways.

The association cortices, as the name implies, interrelate signals arising from the early cortices. They are located everywhere in the cerebral cortex where there are no early sensory cortices or motor cortices. They are organized hierarchically, and the ones higher up in the chain are usually known as higher-order association cortices. The prefrontal cortices and the anterior temporal cortices are examples of higher-order association cortices.

The various regions of the cerebral cortex are traditionally identified by numbers corresponding to the distinctive architectural design of its neuron arrangements, which is known as cytoarchitectonics. The best-known system for numbering the regions was proposed by Brodmann a century ago, and it remains a useful tool today. The Brodmann numbers have nothing whatsoever to do with the area's size or functional importance.

THE IMPORTANCE OF LOCATION

The internal anatomical structure of a brain region is an important determinant of its function. Where a given brain region is located within the three-dimensional volume of a brain is another important determinant. Placement in the global brain volume and internal anatomical structure are largely consequences of evolution, but they are also influenced by individual development. Individual experience shapes the circuitry, and although this influence is most marked at the microcircuitry level, it is inevitably felt at the macroanatomic level as well.

The evolutionary vintage of nuclei is old, a throwback to a time in the history of life when whole brains were little more than chains of ganglia resembling beads in a rosary. A ganglion is, in essence, an individual nucleus before being evolutionarily incorporated into a brain mass. The brains of the nematodes I mentioned in Chapter 2 consist of chains of ganglia.

The location of nuclei within the brain's whole volume is fairly low, always below the mantle provided by the cerebral cortex. They sit in the brain stem, the hypothalamus and thalamus, the basal ganglia, and the basal forebrain (whose extension includes the collection of nuclei known as the amygdalae). Banished as they are from the prime cortical estate, they still have an evolutionary pecking order. The older they are, historically speaking, the closer they are to the brain's midline. And because everything in the brain has two halves, left and right with a dividing median, it so happens that very old nuclei sit looking at their twin on the other side of the midline. This is the case with the brain-stem nuclei that are so vital for life regulation, and for consciousness. In the case of somewhat more modern nuclei—say, the amygdala—the left and right exemplars are more independent and clearly separate from each other.

The cerebral cortices are evolutionarily more recent than the nuclei. They are all distinguished by their two-dimensional sheathlike structure, which confers upon some of them detailed map-making abilities. But the number of layers in a cortex varies from a mere three (for old-vintage cortices) to six (for more recent vintages). The complexity of the circuitry within and across those layers varies as well. The overall location in the whole brain volume is also functionally telling. In general, very modern cortices occur at and around the point at which the major sensory pathways—e.g., auditory, visual, somatosensory—enter the cerebral cortex mantle and are thus connected with sensory processing and map-making. In other words, they belong to the "early sensory cortex" club.

Motor cortices also have varied vintages. Some motor cortices are quite old and small, again located at the midline in the anterior cingulate and supplementary motor regions, clearly visible on the internal (or medial) surface of each cerebral hemisphere. Other motor cortices are modern and structurally sophisticated and occupy a sizable territory on the external surface of the brain (the lateral surface).

What a given region ends up contributing to the overall business of the brain depends significantly on its partners: which talks to the region and which is talked back to, specifically, which regions project their neurons to region X (thus modifying the state of region X) and which regions receive projections from region X (thus being modified by its output). A lot depends on where region X is located within the network. Whether region X has map-making abilities is another important factor in its functional role.

Mind and behavior are the moment-to-moment results of the operation of galaxies of nuclei and cortical parcels articulated by convergent and divergent neural projections. If the galaxies are well organized and work harmoniously, the owner makes poetry. If not, madness ensues.

AT THE INTERFACES BETWEEN THE BRAIN AND THE WORLD

Two kinds of neural structures are located at the border between the brain and the world. One points *inward,* the other *outward.* The first neural structure is made up of the sensory receptors of the body's periphery—the retina, the cochlea in the inner ear, the nerve terminals in our skin, and so forth. These receptors do not receive neuron projections from the outside, at least not naturally, although neuronlike electrical inputs from prosthetic implants are changing this situation. They receive *physical stimuli* instead—light, vibration, mechanical contact.

Sensory receptors initiate a chain of signals from the body's border to inside the brain, across multiple hierarchies of neuron circuits that penetrate deeply into the brain territories. But they don't just move up like water in a pipe system. At every new station they undergo processing and a transformation. In addition, they tend to send signals back to where the inbound projection chains started. These understudied features of brain architecture probably have great significance for certain aspects of consciousness.

The other kind of border point occurs where the *outward* projections from the brain end and the environment begins. The chains of signals arise within the brain but end up either releasing chemical molecules into the atmosphere or connecting to muscular fibers in the body. The latter enables us to move and speak, and that is where the principal outward chains terminate. Beyond the muscle fibers there lies direct movement in space. In earlier stages of evolution, the release of chemical molecules at the membrane or skin border played important roles in the life of an organism. It was an important means of action. In humans, this facet remains understudied, although the release of pheromones is not in doubt.

One may conceptualize the brain as a progressive elaboration of what began as a simple reflex arc: neuron NEU senses object OB and signals to neuron ZADIG, which projects to muscle fiber MUSC and causes movement. Later in evolution a neuron would be added to the reflex circuit, in between NEU and ZADIG. This is an *interneuron,* and let us call it INT; it behaves such that the response of neuron ZADIG is no longer automatic. Neuron ZADIG responds, for example, only if neuron NEU fires all its guns upon it and not if neuron ZADIG receives a weaker message; a critical part of the decision is left in the hands of the interneuron INT.

A major aspect of brain evolution has consisted of adding the equivalent of interneurons at every level of brain circuitry—a slew of such equivalents, in fact. The largest such equivalents, located in the cerebral cortex, might well be called *interregions*. They become sandwiched between other regions, for the good and obvious purpose of modulating simple responses to varied stimuli and making the responses less simple, less automatic.

On the path to making the modulation more subtle and sophisticated, the brain developed systems that map stimuli in such detail that the ultimate consequence was images and mind. Eventually the brain added a self process to those minds, and that permitted the creation of novel responses. Finally, in humans, when such conscious minds were organized in collectives of like beings, the creation of cultures became possible along with their attending external artifacts. In turn, cultures have influenced the operation of brains over generations and eventually influenced the evolution of the human brain.

The brain is a system of systems. Each system is composed of an elaborate interconnection of small but macroscopic cortical regions and subcortical nuclei, which are made of microscopic local circuits, which are made of neurons, all of which are connected by synapses.

What neurons do depends on the local assembly of neurons to which they belong; what systems end up doing depends on how local assemblies influence other assemblies within an interconnected architecture; finally, whatever each assembly contributes to the function of the system to which it belongs depends on its place in that system.

A NOTE ON THE MIND-BRAIN EQUIVALENCE HYPOTHESIS

The perspective adopted in this book contains a hypothesis that is not universally liked, let alone accepted—namely, the idea that mental states and brain states are essentially equivalent. The reasons for the reluctance in endorsing such a hypothesis deserve a hearing.

In the physical world, of which the brain is unequivocally a part, equivalence and identity are defined by physical attributes such as mass, dimensions, movement, charge, and so forth. Those who reject the identity between physical states and mental states suggest that while a brain map that corresponds to a particular physical object can be discussed in physical terms, it would be absurd to discuss the respective mental pattern in physical terms. The reason given is that to date science has not been able to determine the physical attributes of mental patterns, and if science cannot do so, then the mental cannot be identified with the physical. I fear, however, that this reasoning may not be sound. Let me explain why I think so.

First, we need to consider how we determine that nonmental states are physical. In the case of objects out in the world, we proceed by perceiving them with our peripheral sensory probes and by using varied instruments to execute measurements. In the case of mental events, however, we cannot do the same. This is not because mental events are not equivalent to neural states but because, given their place of occurrence—the interior of the brain—mental states are simply not available for measurement. In fact, mental events can be perceived only by part of the very same process that includes them—the mind, that is. The situation is unfortunate but says nothing whatsoever about the physicality of the mind or lack thereof. The situation does impose major qualifications on the intuitions that can emerge from it, however, and it is thus prudent to doubt the traditional view that asserts that mental states *cannot* be equivalent to physical states. It is unreasonable to endorse such a view purely on the basis of introspective observations. The personal perspective should be used and enjoyed for what it gives us directly: experience that can be made conscious, and that can help guide our life, provided extensive reflective analysis conducted offline—which includes scientific scrutiny—validates its counsel.

The fact that neural maps and the corresponding images are found *inside* the brain, accessible only to the brain's owner, is a hurdle. But where else would the maps/images be found but within a private, secluded sector of the brain, given that they are formed inside the brain to begin with? What would be surprising would be to find them outside the brain, given that brain anatomy is not designed to externalize them.

For the time being, the mental state/brain state equivalence should be regarded as a useful hypothesis rather than a certainty. It will take a continued accrual of evidence to lend it support, and for that we need an additional perspective, informed by evidence from evolutionary neurobiology aligned with varied neuroscience evidence.

Some may question the need for an additional perspective to make sense of mental events, but there are good justifications for an added perspective. The facts that mental events are *correlated* with brain events—and no one disputes that fact—and that the latter exist inside the brain, inaccessible to direct measurement, justify a special approach. Also, given that mental/brain events are certainly the product of a long history of biological evolution, it makes sense that evolutionary evidence be included in their consideration. Last, given that mental/brain events are possibly the most complex phenomena in nature, the need for special treatment should not be regarded as exceptional.

Even with the help of neuroscience techniques more powerful than are available today, we are unlikely ever to chart the full scope of neural phenomena associated with a mental state, even a simple one. What is possible and needed, for the time being, is a gradual theoretical approximation supported by new empirical evidence.

Accepting the hypothesized mental/neural equivalence is especially helpful with the vexing problem of downward causality. Mental states do exert their influence on behavior, as can be easily revealed by all manner of actions executed by the nervous system and the muscles at its command. The problem, some will say the mystery, has to do with how a phenomenon that is regarded as nonphysical—the mind—can exert its influence on the very physical nervous system that moves us to action. Once mental states and neural states are regarded as the two faces of the same process, one more Janus out to trick us, downward causality is less of a problem.

On the other hand, rejecting mind/brain equivalence requires a problematic assumption: that somehow it would be less natural and plausible for neurons to create mappings of things, and for these mappings to be fully formed mental events, than it is for other cells in the organism to create, for example, the shapes of body parts or to execute body actions. When cells in the body proper are placed together in a particular spatial configuration, according to a plan, they constitute an object.

A hand is a good example. It is made of bones, muscles, tendons, connective tissue, a network of blood vessels and another of nerve pathways, and several layers of skin, all put into place according to a specific architectural pattern. When such a biological object moves in space, it performs an action, for example, your hand pointing to me. Both object and action are physical events in space and time. Now, when neurons arranged in a two-dimensional sheath are active or inactive according to the inputs they receive, they create a pattern. When the pattern corresponds to some object or action, it constitutes a map of something else, a map of that object or that action. Grounded as it is in the activity of physical cells, the pattern is just as physical as the objects or actions it corresponds to. The pattern is momentarily *drawn* in the brain, *carved* in the brain by its activity. Why would circuits of brain cells not create some sort of imagetic correspondence for things, provided the cells are properly wired, operate as they are supposed to operate, and become active when they should? Why would the resulting momentary activity patterns necessarily be any less physical than the objects and actions were in the first place?

NOTES

1 / Awakening

1. I became aware of the opposition to consciousness research in the late 1980s, when I first talked about the issue with Francis Crick. By then Francis was thinking of putting aside his favorite neuroscience topics and turning his efforts toward consciousness. I was not quite ready to do the same, a wise move given the mood of the times. I remember Francis asking me, with characteristic amusement, if I knew Stuart Sutherland's definition of *consciousness*. I did not. Sutherland, a British psychologist famous for his dismissive and devastating remarks about varied issues and colleagues, had just published in his *Dictionary of Psychology* a startling definition that Francis proceeded to read: "Consciousness is a fascinating but elusive phenomenon; it is impossible to specify what it is, what it does, or why it evolved. Nothing worth reading has been written about it." Stuart Sutherland, *International Dictionary of Psychology,* 2nd ed. (New York: Continuum, 1996).

We laughed heartily, and before we considered the merits of this masterpiece of enthusiasm, Francis read me Sutherland's definition of *love*. Here it is, for the curious reader: "A form of mental illness not yet recognized by any of the standard diagnostic manuals." We laughed some more.

Even by the standards of the day, Sutherland's statement was extreme, although it did capture a widely held attitude: the time for consciousness research, by which everyone really meant research on how the brain accounts for consciousness, had not yet come. The attitude did not paralyze the field, but in retrospect it was pernicious: it artificially separated the consciousness problem from the mind problem. It certainly gave neuroscientists license to continue investigating the mind without having to confront the hurdles posed by the study of consciousness. (Surprisingly, I met Sutherland many years later and told him what I was up to on the issues of mind and self. He seemed to like the ideas and was extremely kind to me.)

The negative attitude is by no means gone. I respect the skepticism of the colleagues who still hold it, but the idea that explaining the emergence of conscious minds is beyond current intelligence strikes me as very odd and probably false, as does the idea that we must wait for the next Darwin or Einstein to solve the mystery. The same intelligence that, for example, can ambitiously tackle the evolutionary history of biology and uncover the genetic coding behind our lives ought to at least try to address the problem of consciousness before declaring defeat. Darwin, by the way, did not think that consciousness was the Everest of science, and I sympathize with that view.

As for Einstein, who looked at nature through Spinoza's lenses, it is difficult to imagine consciousness fazing him, had the notion of elucidating it ever come under his purview.

2. Beginning about a decade ago, in scientific articles and in a book, I specifically addressed the problem of consciousness. See Antonio Damasio, "Investigating the Biology of Consciousness," *Philosophical Transactions of the Royal Society B: Biological Sciences* 353 (1998); Antonio Damasio, *The Feeling of What Happens: Body and Emotion in the Making of Consciousness* (New York: Harcourt Brace, 1999); Josef Parvizi and Antonio Damasio, "Consciousness and the Brainstem," *Cognition* 79 (2001), 135–59; Antonio Damasio, "The Person Within," *Nature* 423 (2003), 227; Josef Parvizi and Antonio Damasio, "Neuroanatomical Correlates of Brainstem Coma," *Brain* 126 (2003), 1524–36; David Rudrauf and A. R. Damasio, "A Conjecture Regarding the Biological Mechanism of Subjectivity and Feeling," *Journal of Consciousness Studies* 12 (2005), 236–62; Antonio Damasio and Kaspar Meyer, "Consciousness: An Overview of the Phenomenon and of Its Possible Neural Basis," in *The Neurology of Consciousness: Neuroscience and Neuropathology,* ed. Steven Laureys and Giulio Tononi (London: Academic Press, 2009).

3. W. Penfield, "Epileptic Automatisms and the Centrencephalic Integrating System," *Research Publications of the Association for Nervous and Mental Disease* 30 (1952), 513–28; W. Penfield and H. H. Jasper, *Epilepsy and the Functional Anatomy of the Human Brain* (New York: Little, Brown, 1954); G. Moruzzi and H. W. Magoun, "Brain Stem Reticular Formation and Activation of the EEG," *Electroencephalography and Clinical Neurophysiology* 1, no. 4 (1949), 455–73.

4. For a review of the relevant literature, I recommend the current edition of a classic: Jerome B. Posner, Clifford B. Saper, Nicholas D. Schiff, and Fred Plum, *Plum and Posner's Diagnosis of Stupor and Coma* (New York: Oxford University Press, 2007).

5. William James, *The Principles of Psychology* (New York: Dover Press, 1890).

6. A "hint half guessed" and a "gift half understood" are words I borrowed from T. S. Eliot to describe this elusiveness in Damasio, *Feeling of What Happens*.

7. James, *Principles,* 1, chap. 2.

8. A. Damasio, "The Somatic Marker Hypothesis and the Possible Function of the Prefrontal Cortex," *Philosophical Transactions of the Royal Society B: Biological Sciences* 351, no. 1346 (1996), 1413–20; A. Damasio, *Descartes' Error* (New York: Putnam, 1994).

9. John Searle, *The Mystery of Consciousness* (New York: New York Review Books, 1990).

10. Preferring to approach consciousness through perception and deferring interest in the self has been a standard strategy, exemplified by Francis Crick and Christof Koch in "A Framework for Consciousness," *Nature Neuroscience* 6, no. 2 (2003), 119–26. A notable exception, contained in a volume that deals mostly with emotion, is J. Panksepp, *Affective Neuroscience: The Foundation of Human and Animal Emotions* (New

York: Oxford University Press, 1998). Rodolfo Llinás also acknowledges the importance of the self; see his *I of the Vortex: From Neurons to Self* (Cambridge, Mass.: MIT
Press, 2002). Gerald Edelman's thinking on consciousness implies the presence of a self
process, although that is not the focus of his proposals in *The Remembered Present: A
Biological Theory of Consciousness* (New York: Basic Books, 1989).

11. The gist of the disagreement is discussed in James, *Principles,* 1, 350–52. Hume's
assertion and James's response are as follows:

HUME: "For my part, when I enter most intimately into what I call *myself,* I
always stumble on some particular perception or other of heat or cold, light or
shade, love or hatred, pain or pleasure. I never can catch *myself* at any time
without a perception, and never can observe anything but the perception.
When my perceptions are removed for any time, as by sound sleep, so long am
I insensible of *myself,* and may truly be said not to exist. And were all my perceptions removed by death, and could I neither think, nor feel, nor see, nor
love, nor hate after the dissolution of my body, I should be entirely annihi-
lated, nor do I conceive what is farther requisite to make me a perfect non-
entity. If anyone, upon serious and unprejudiced reflection, thinks he has a
different notion of *himself,* I must confess I can reason no longer with him. All
I can allow him is, that he may be in the right as well as I, and that we are essentially different in this particular. He may, perhaps, perceive something simple
and continued which he calls *himself;* though I am certain there is no such principle in me." Hume, *Treatise on Human Nature,* book 1.

JAMES: "But Hume, after doing this good piece of introspective work, pro-
ceeds to pour out the child in the bath, and to fly to as great an extreme as the
substantialist philosophers. As they say the Self is nothing but Unity, unity
abstract and absolute, so Hume says it is nothing but Diversity, diversity
abstract and absolute; whereas in truth it is that mixture of unity and diversity
which we ourselves have already found so easy to pick apart . . . he denies this
thread of resemblance, this core of sameness running through the ingredients
of the Self, to exist even as a phenomenal thing."

12. D. Dennet, *Consciousness Explained* (New York: Little, Brown, 1992); S. Gallagher, "Philosophical Conceptions of Self: Implications for Cognitive Science,"
Trends in Cognitive Science 4, no. 1 (2000), 14–21; G. Strawson, "The Self," *Journal of
Consciousness Studies* 4, nos. 5–6 (1997), 405–28. In addition to the works cited in note
10, see also Damasio, *Feeling of What Happens;* P. S. Churchland, "Self-Representation
in Nervous Systems," *Science* 296, no. 5566 (2002), 308–10; J. LeDoux, *The Synaptic Self:
How Our Brains Become Who We Are* (New York: Viking Press, 2002); Chris Frith,
Making Up the Mind: How the Brain Creates Our Mental World (New York: Wiley-
Blackwell, 2007); G. Northoff, A. Heinzel, M. de Greck, F. Bermpohl, H.
Doborowolny, and J. Panksepp, "Self-referential Processing in Our Brain—A Meta-
analysis of Imaging Studies on the Self," *NeuroImage* 31, no. 1 (2006), 440–57.

13. The work of Roger Penrose and Stuart Hameroff exemplifies this position, which has also been championed by the philosopher David Chalmers. See R. Penrose, *The Emperor's New Mind: Concerning Computers, Minds, and the Laws of Physics* (Oxford: Oxford University Press, 1989); S. Hameroff, "Quantum Computation in Brain Microtubules? The Penrose-Hameroff 'Orch OR' Model of Consciousness," *Philosophical Transactions of the Royal Society A: Mathematical, Physical and Engineering Sciencies* 356 (1998), 1869–96; David Chalmers, *The Conscious Mind: In Search of a Fundamental Theory* (Oxford: Oxford University Press, 1996). The point about the coincidence of mysteries was argued convincingly in Patricia S. Churchland and Rick Grush, "Computation and the Brain," in *The MIT Encyclopedia of Cognitive Science*, ed. R. Wilson (Cambridge, Mass.: MIT Press, 1998).

14. The false intuition is strengthened by the claim that the dimensions or mass of mental states cannot be measured with conventional instruments. That is undeniably true, but the situation is a consequence of the location of mental events (the recondite interior of the brain) where conventional measurements are not possible. The situation is frustrating for observers, but it says nothing about the physicality, or lack thereof, of mental states. States of mind begin physically, and physical they remain. They can be revealed only when an equally physical construction called self becomes available and does its witnessing job. The traditional conceptions of *matter* and *mental* are unnecessarily narrow. The burden of proof does rest with those who find it natural for mind states to be constituted by brain activity. But endorsing the intuitive mind-brain split as the only platform for discussing the problem is not likely to encourage the search for additional proof.

15. Evolutionary thinking is also a major factor in the consciousness proposals of, among others, Gerald Edelman, Jaak Panksepp, and Rodolfo Llinás. See also Nicholas Humphrey, *Seeing Red: A Study in Consciousness* (Cambridge, Mass.: Harvard University Press, 2006). For examples of evolutionary thinking applied to the understanding of the human mind, see E. O. Wilson (a pioneer in the field), *Consilience: The Unity of Knowledge* (New York: Knopf, 1998), and Steven Pinker, *How the Mind Works* (New York: Norton, 1997).

16. For fundamental work on selectional pressures in individual brain development, see Jean-Pierre Changeux, *Neuronal Man: The Biology of Mind* (New York: Pantheon, 1985), and Edelman, *Remembered Present*.

17. My previous accounts of the self did not include the primordial self. The elementary feeling of existence was part of the core self. I came to the conclusion that the process can work only if the brain-stem component of the protoself generates an elementary feeling, a primitive of sorts, independently of any object interacting with the organism and thus modifying the protoself. Jaak Panksepp has long championed a somewhat comparable view of the process and has also given it a brain-stem origin. See Panksepp, *Affective Neuroscience*. Panksepp's views differ in the following ways. First, the simple feeling that he posits appears to be necessarily related to external events in the world. He describes it as "that ineffable feeling of experiencing oneself as

an active agent in the perceived events of the world." On the other hand, the primitive feeling/primordial self, I propose, is a spontaneous product of the protoself. In theory, primordial feelings occur regardless of whether the protoself is engaged by objects and events external to the brain. They need to be related to the living body and nothing else. Panksepp's description matches more closely my description of the core self, which does include a feeling of knowing relative to an object. It appears to be a notch up in the construction scale. Second, Panksepp relates this primary consciousness mainly to motor activities in structures of the brain stem (periaqueductal gray, cerebellum, superior colliculi), while I place the emphasis in sensory structures such as nucleus tractus solitarius and parabrachial nucleus, albeit in close association with the periaqueductal gray and deep layers of superior colliculi.

18. The study of the links between neurobiological networks, on the one hand, and social networks, on the other, is an important area of investigation. See Manuel Castells, *Communication Power* (New York: Oxford University Press, 2009).

19. See F. Scott Fitzgerald, *The Diamond as Big as the Ritz* (New York: Scribner's, 1922).

2 / From Life Regulation to Biological Value

1. Some of the sources for the concepts discussed in this section are as follows: Gerald M. Edelman, *Topobiology: An Introduction to Molecular Embryology* (New York: Basic Books, 1988); Christian De Duve, *Blueprint for a Cell: The Nature and Origin of Life* (Burlington, N.C.: Neil Patterson, 1991); Robert D. Barnes and Edward E. Ruppert, *Invertebrate Zoology* (New York: Saunders College Publishing, 1994); Eshel Ben-Jacob, Ofer Schochet, Adam Tenenbaum, Inon Cohen, Andras Czirók, and Tamas Vicsek, "Generic Modeling of Cooperative Growth Patterns in Bacterial Colonies," *Nature* 368, no. 6466 (1994), 46–49; Christian De Duve, *Vital Dust: Life as a Cosmic Imperative* (New York: Basic Books, 1995); Ann B. Butler and William Hodos, *Comparative Vertebrate Neuroanatomy* (Hoboken, N. J.: Wiley Interscience, 2005); Andrew H. Knoll, *Life on a Young Planet* (Princeton, N.J.: Princeton University Press, 2003); Bert Holldobler and Edward O. Wilson, *The Superorganism* (New York: W.W. Norton, 2009); Jonathan Flint, Ralph J. Greenspan, and Kenneth Kendler, *How Genes Influence Behavior* (New York: Oxford University Press, 2010).

2. Lynn Margulis, *Symbiosis in Cell Evolution: Microbial Communities* (San Francisco: W. H. Freeman, 1993); L. Sagan, "On the Origin of Mitosing Cells," *Journal of Theoretical Biology* 14 (1967), 225–74; J. Shapiro, "Bacteria as Multicellular Organisms," *Scientific American* 256, no. 6 (1998), 84–89.

3. In previous writings I have alluded to this behavioral anticipation and preview, in simple organisms, of attitudes that we usually associate with complex human behavior. See Antonio Damasio, *The Feeling of What Happens: Body and Emotion in the Making of Consciousness* (New York: Harcourt Brace, 1999); and *Looking for Spinoza* (New York: Harcourt Brace, 2003). Rodolfo Llinás makes comparable comments in *I*

of the Vortex: From Neurons to Self (Cambridge, Mass.: MIT Press, 2002), as does T. Fitch in "Nano-intentionality: A Defense of Intrinsic Intentionality," *Biology and Philosophy,* 23, no. 2 (2007), 157–77.

4. For a review on the general physiology of neurons, see Eric R. Kandel, James H. Schwartz, and Thomas M. Jessel, *Principles of Neural Science,* 4th ed. (New York: McGraw-Hill, 2000).

5. De Duve, *Vital Dust*.

6. Claude Bernard, *An Introduction to the Study of Experimental Medicine* (1865), trans. Henry Copley Greene (New York: Macmillan, 1927); Walter Cannon, *The Wisdom of the Body* (New York: W. W. Norton, 1932).

7. Answers regarding the origins of homeostasis need to be found at even simpler levels. The behavior of certain molecules is behind their spontaneous assembly in arrangements such as RNA and DNA. Here we are confronting questions about the very origin of life. We can say with some confidence that the conformation of some molecules lends them a natural "self" preservation, as close to the first light of homeostasis as one can get at the moment.

8. For a review of neuroscience on the notion of value, see Read Montague, *Why Choose This Book?: How We Make Decisions* (London: Penguin, 2006). A recent volume on decision-making devotes attention to the notion of value: Paul W. Glimcher et al., eds., *Neuroeconomics: Decision Making and the Brain* (London: Academic Press, 2009), especially Peter Dayan and Ben Seymour, "Values and Actions in Aversion"; Antonio Damasio, "Neuroscience and the Emergence of Neuroeconomics"; Wolfram Schultz, "Midbrain Dopamine Neurons: A Retina of the Reward System?"; Bernard W. Balleine, Nathaniel D. Daw, and John P. O'Doherty, "Multiple Forms of Value Learning and the Function of Dopamine"; Brian Knutson, Mauricio R. Delgado, and Paul E. M. Phillips, "Representation of Subjective Value in the Striatum"; and Kenji Doya and Minoru Kimura, "The Basal Ganglia and Encoding of Value."

9. For a clear picture of the complexity of homeostatic regulation, see Alan G. Watts and Casey M. Donovan, "Sweet Talk in the Brain: Glucosensing, Neural Networks, and Hypoglycemic Counterregulation," *Frontiers in Neuroendocrinology* 31 (2010), 32–43.

10. C. Bargmann, "Olfaction—From the Nose to the Brain," *Nature* 384, no. 6609 (1996), 512–13; C. Bargmann, "Neuroscience: Comraderie and Nostalgia in Nematodes," *Current Biology* 15 (2005), R832–33. I thank Baruch Blumberg for alerting me to the concept of "quorum sensing."

11. The automated, nonminded, and nonconscious life regulation of simple organisms is good enough to permit survival in environments that offer abundant nutrients and a low risk of conditions such as temperature variations or the presence of predators. But such simple organisms must remain within the environments to which they are adapted or face extinction. Most species still in existence do very well indeed in their ecological niche and operate under automated life regulation *only*.

Moving out of the ecological niche opens up all sorts of possibilities for the roving, trespassing creature. But trespassing comes at a potential cost. In situations of scarcity, survival is possible only when the trespasser is equipped with sophisticated devices that allow it new options of behavior. These new devices must offer valuable "advice" in the form of making the trespasser go elsewhere to find what it needs, and they must suggest alternate, safe means of doing so. The new devices also allow the trespasser to predict incoming risks, such as predators, and provide a means to elude them.

3 / Making Maps and Making Images

1. Rodolfo Llinás, cited earlier.

2. For a clear review of why the brain is not a blank slate, see Steve Pinker, *The Blank State: The Modern Denial of Human Nature* (New York: Viking, 2002).

3. R. B. H. Tootell, E. Switkes, M. S. Silverman, et al., "Functional Anatomy of Macaque Striate Cortex. II. Retinotopic Organization," *Journal of Neuroscience* 8 (1983), 1531–68; K. Meyer, J. T. Kaplan, R. Essex, C. Webber, H. Damasio and A. Damasio, "Predicting Visual Stimuli on the Basis of Activity in Auditory Cortices," *Nature Neuroscience* 13 (2010), 667–668; G. Rees and J. D. Haynes, "Decoding Mental States from Brain Activity in Humans," *Nature Reviews Neuroscience* 7 (July 7, 2006), 523–34. Also see Gerald Edelman, *Neural Darwinism: The Theory of Neuronal Group Selection* (New York: Basic Books, 1987), for a valuable discussion of neural maps and for his insistence on the notion of value applied to the selection of maps; David Hubel and Torsten Wiesel, *Brain and Visual Perception* (New York: Oxford University Press, 2004).

4. The stamping of value is possibly made on the basis of an emotional marker, a somatic marker, as I have proposed elsewhere: A. Damasio, "The Somatic Marker Hypothesis and the Possible Functions of the Prefrontal Cortex," *Philosophical Transactions of the Royal Society B: Biological Sciences* 351 (1996), 1413–20.

5. For reviews of pertinent neuropsychology literature, see H. Damasio and A. Damasio, *Lesion Analysis in Neuropsychology* (New York: Oxford University Press, 1989); Kenneth M. Heilman and Edward Valenstein, eds., *Clinical Neuropsychology*, 4th ed. (Oxford: Oxford University Press, 2003); H. Damasio and A. R. Damasio, "The Neural Basis for Memory, Language and Behavioral Guidance: Advances with the Lesion Method in Humans," *Seminars in the Neurosciences* 2 (1990), 277–96; A. Damasio, D. Tranel, and M. Rizzo, "Disorders of Complex Visual Processing," in *Principles of Behavioral and Cognitive Neurology,* ed. M. M. Mesulam (New York: Oxford University Press, 2000).

6. Bjorn Merker is another author who has argued for the brain stem as an origin for the mind and even for consciousness in "Consciousness Without a Cerebral Cortex," *Behavioral and Brain Sciences* 30 (2007), 63–81.

7. Antonio R. Damasio, Paul J. Eslinger, Hanna Damasio, Gary W. Van Hoesen, and Steven Cornell, "Multimodal Amnesic Syndrome Following Bilateral Temporal

and Basal Forebrain Damage," *Archives of Neurology* 42, no. 3 (1985), 252–59; Justin S. Feinstein, David Rudrauf, Sahib S. Khlasa, Martin D. Cassell, Joel Bruss, Thomas J. Grabowski, and Daniel Tranel, "Bilateral Limbic System Destruction in Man," *Journal of Clinical and Experimental Neuropsychology,* September 17, 2009, 1–19.

8. It might be countered that in the absence of the insula, other somatosensory cortices (SI, SII) might provide the source for feelings; or that the anterior cingulate cortices might as well, since they are often active in emotional-feeling studies using fMRI. This idea is problematic on several counts. First, the anterior cingulate cortices are essentially motor structures, involved in creating emotional responses rather than sensing them. Second, visceral information is first channeled to the insula and only then distributed to SI and SII. Extensive damage to the insula precludes this process. Third, fMRI studies of bodily and emotional feelings in normal individuals reveal systematic and abundant insular activations but rare activations in SI and SII, a finding that is in line with the fact that SI and SII are dedicated to exteroception and proprioception (the mapping of touch, pressure, and skeletal movement) rather than interoception (the mapping of the viscera and internal milieu). In fact, pain of visceral origin tends not to map well onto SI, as shown by M. C. Bushnell, G. H. Duncan, R. K. Hofbauer, B. Ha, J.-I.- Chen, and B. Carrier, "Pain Perception: Is There a Role for Primary Somatosensory Cortex?" *Proceedings of the National Academy of Sciences* 96 (1999), 7705–09.

9. J. Parvizi and A. R. Damasio, "Consciousness and the Brainstem," *Cognition* 79 (2001), 135–60.

10. Alan D. Shewmon, Gregory L. Holmes, and Paul A. Byrne, "Consciousness in Congenitally Decorticate Children: Developmental Vegetative State as a Self-fulfilling Prophecy," *Developmental Medicine and Child Neurology* 41 (1999), 364–74.

11. Bernard M. Strehler, "Where Is the Self? A Neuroanatomical Theory of Consciousness," *Synapse* 7 (1991), 44–91; J. Panksepp, *Affective Neuroscience: The Foundation of Human and Animal Emotions* (New York: Oxford University Press, 1998). See also Merker, "Consciousness."

12. The mapped arrangement of the retina is preserved, and the activity of the left colliculus goes with the right visual field and vice versa. The neurons in the superficial layers of the superior colliculus prefer to respond to moving rather than stationary stimuli, and to slow-moving rather than fast-moving stimuli. They also prefer stimuli that move across the visual field in a specific direction. The vision provided by the superior colliculus privileges the detection and tracking of moving targets.

Unlike the superficial layers, the deep layers of the colliculus are connected to a variety of structures related to vision, hearing, body sensation, and movement. The visual input reaches these layers directly from the contralateral retina. The auditory input reaches them from the inferior colliculus. The somatosensory input arrives from the spinal cord, the trigeminal nucleus, the vagal nucleus, the area postrema, and the hypothalamus. Proprioceptive information, the variety of somatosensory information having to do with the musculature, reaches the superior colliculus from the spinal cord via the cerebellum. Vestibular information is transmitted via projections via the fastigial nucleus.

13. The contrast between the superior and inferior colliculus is quite suggestive. The inferior colliculus is also a layered structure, but its domain is purely auditory. It is an important way station for auditory signals en route to the cerebral cortex. The superior colliculus has both a visual domain, tied to its superficial layers, and a coordinating domain, linked to the deep layers. See Paul J. May, "The Mammalian Superior Colliculus: Laminar Structure and Connections," *Progress in Brain Research* 151 (2006), 321–78; Barry E. Stein, "Development of the Superior Colliculus," *Annual Review of Neuroscience* 7 (1984), 95–125; Eliana M. Klier, Hongying Wang, and Douglas J. Crawford, "The Superior Colliculus Encodes Gaze Commands in Retinal Coordinates," *Nature Neuroscience* 4, no. 6 (2001), 627–32; and Michael F. Huerta and John K. Harting, "Connectional Organization of the Superior Colliculus," *Trends in Neurosciences,* August 1984, 286–89.

14. Bernard M. Strehler, "Where Is the Self? A Neuroanatomical Theory of Consciousness," *Synapse* 7 (1991), 44–91; Merker, "Consciousness."

15. D. Denny Brown, "The Midbrain and Motor Integration," *Proceedings of the Royal Society of Medicine* 55 (1962), 527–38.

16. Michael Brecht, Wolf Singer, and Andreas K. Engel, "Patterns of Synchronization in the Superior Colliculus of Anesthetized Cats," *Journal of Neuroscience* 19, no. 9 (1999), 3567–79; Michael Brecht, Rainer Goebel, Wolf Singer, and Andreas K. Engel, "Synchronization of Visual Responses in the Superior Colliculus of Awake Cats," *NeuroReport* 12, no. 1 (2001), 43–47; Michael Brecht, Wolf Singer, and Andreas K. Engel, "Correlation Analysis of Corticotectal Interactions in the Cat Visual System," *Journal of Neurophysiology* 79 (1998), 2394–407.

17. W. Singer, "Formation of Cortical Cell Assemblies," *Symposium on Qualitative Biology* 55 (1990), 939–52; Llinás, *I of the Vortex.*

18. L. Melloni, C. Molina, M. Pena, D. Torres, W. Singer, and E. Rodríguez, "Synchronization of Neural Activity Across Cortical Areas Correlates with Conscious Perception," *Journal of Neuroscience* 27, no. 11 (2007), 2858–65.

4 / The Body in Mind

1. Franz Brentano, *Psychology from an Empirical Standpoint,* trans. Antos C. Rancurello, D. B. Terrel, and Linda L. McAllister (London: Routledge, 1995), 88–89.

2. Daniel Dennett, *The Intentional Stance* (Cambridge, Mass.: MIT Press, 1987), has long made this same argument, and recently so has Tecumseh Fitch in "Nano-intentionality: A Defense of Intrinsic Intentionality," *Biology and Philosophy* 23, no. 2 (2007), 157–77.

3. William James, *The Principles of Psychology* (New York: Dover Press, 1890). James's treatment of the body as relevant to the understanding of the mind was largely

neglected in neuroscience until quite recently. In philosophy, however, the body continued to play a central role, a prominent example being Maurice Merleau-Ponty, *Phenomenology of Perception* (London: Routledge, 1962). Among contemporary philosophers, Mark Johnson is the recognized leader in this area. The body played a prominent role in his well-known work with George Lakoff, *Metaphors We Live By* (Chicago: University of Chicago Press, 1980), but two later monographs are definitive treatments of the topic: Mark Johnson, *The Body in the Mind: The Bodily Basis of Meaning, Imagination, and Reason* (Chicago: University of Chicago Press, 1987); and Mark Johnson, *The Meaning of the Body: Aesthetics of Human Understanding* (Chicago: University of Chicago Press, 2007).

4. Julian Jaynes, *The Origin of Consciousness in the Breakdown of the Bicameral Mind* (New York: Houghton Mifflin, 1976).

5. The two critical figures in this history are Ernst Heinrich Weber and Charles Scott Sherrington. See Weber, *Handwörterbuch des Physiologie mit Rücksicht auf physiologische Pathologie,* ed. R. Wagner (Braunschwieg, Germany: Biewig und Sohn, 1846), and Sherrington, *Text-book of Physiology,* ed. E. A. Schäfer (Edinburgh: Pentland, 1900). Regrettably, by the time he revised his famous textbook, Sherrington had abandoned the German concept of general bodily feeling or *Gemeingefühl* and no longer emphasized his early notion of "material me." See C. S. Sherrington, *The Integrative Action of the Nervous System* (Cambridge: Cambridge University Press, 1948). A. D. Craig provides an accurate historical review of this state of affairs in "How Do You Feel? Interoception: The Sense of the Physiological Condition of the Body," *Nature Reviews Neuroscience* 3 (2002), 655–66.

6. The basics of the body-to-brain interconnection are well reviewed in Clifford Saper, "The Central Autonomic Nervous System: Conscious Visceral Perception and Autonomic Pattern Generation," *Annual Review of Neuroscience* 25 (2002), 433–69. See also Stephen W. Porges, "The Polyvagal Perspective," *Biological Psychology* 74 (2007), 116–43. The structure of the brain stem and hypothalamic nuclei in charge of executing this two-way process can be gleaned from the following articles: Caroline Gauriau and Jean-François Bernard, "Pain Pathways and Parabrachial Circuits in the Rat," *Experimental Physiology* 87, no. 2 (2001), 251–58; M. Giola, R. Luigi, Maria Grazia Pretruccioli, and Rossella Bianchi, "The Cytoarchitecture of the Adult Human Parabrachial Nucleus: A Nissl and Golgi Study," *Archives of Histology and Cytology* 63, no. 5 (2001), 411–24; Michael M. Behbahani, "Functional Characteristics of the Midbrain Periaqueductal Gray," *Progress in Neurobiology* 46 (1995), 575–605; Thomas M. Hyde and Richard R. Miselis, "Subnuclear Organization of the Human Caudal Nucleus of the Solitary Tract," *Brain Research Bulletin* 29 (1992), 95–109; Deborah A. McRitchie and Istvan Törk, "The Internal Organization of the Human Solitary Nucleus," *Brain Research Bulletin* 31 (1992), 171–93; Christine H. Block and Melinda L. Estes, "The Cytoarchitectural Organization of the Human Parabrachial Nuclear Complex," *Brain Research Bulletin* 24 (1989), 617–26; L. Bourgeais, L. Monconduit, L. Villanueva, and J. F. Bernard, "Parabrachial Internal Lateral Neurons Convey Noci-

ceptive Messages from the Deep Laminas of the Dorsal Horn to the Intralaminar Thalamus," *Journal of Neuroscience* 21 (2001), 2159–65.

7. A. Damasio, *Descartes' Error* (New York: Putnam, 1994).

8. M. E. Goldberg and C. J. Bruce, "Primate Frontal Eye Fields. III. Maintenance of a Spatially Accurate Saccade Signal," *Journal of Neurophysiology* 64 (1990), 489–508; M. E. Goldberg and R. H. Wurtz, "Extraretinal Influences on the Visual Control of Eye Movement," in *Motor Control: Concepts and Issues,* ed. D. R. Humphrey and H.-J. Freund (Chichester, U.K.: Wiley, 1991), 163–79.

9. G. Rizzolatti and L. Craighero, "The Mirror-Neuron System," *Annual Review of Neuroscience* 27 (2004), 169–92; V. Gallese, "The Shared Manifold Hypothesis," *Journal of Consciousness Studies* 8 (2001), 33–50.

10. R. Hari, N. Forss, S. Avikainen, E. Kirveskari, S. Salenius, and G. Rizzolatti, "Activation of Human Primary Motor Cortex During Action Observation: A Neuromagnetic Study," *Proceedings of the National Academy of Science* 95 (1998), 15061–65.

11. Tania Singer, Ben Seymour, John O'Doherty, Holger Kaube, Raymond J. Dolan, and Chris D. Frith, "Empathy for Pain Involves the Affective but Not Sensory Components for Pain," *Science* 303 (2004), 1157–62.

12. R. Adolphs, H. Damasio, D. Tranel, G. Cooper, and A. Damasio, "A Role for Somatosensory Cortices in the Visual Recognition of Emotion as Revealed by Three-Dimensional Lesion Mapping," *Journal of Neuroscience* 20 (2000), 2683–90.

5 / Emotions and Feelings

1. Martha C. Nussbaum, *Upheavals of Thought: The Intelligence of Emotions* (Cambridge: Cambridge University Press, 2001).

2. R. M. Sapolsky, *Why Zebras Don't Get Ulcers: An Updated Guide to Stress, Stress-related Diseases, and Coping* (New York: W. H. Freeman, 1998); David Servan-Schreiber, *The Instinct to Heal: Curing Stress, Anxiety, and Depression Without Drugs and Without Talk Therapy* (Emmaus, Pa.: Rodale, 2004).

3. William James, "What Is an Emotion?" *Mind* 9 (1884), 188–205.

4. W. B. Cannon, "The James-Lange Theory of Emotions: A Critical Examination and an Alternative Theory," *American Journal of Psychology* 39 (1927), 106–24.

5. Antonio Damasio, *Descartes' Error* (New York: Putnam, 1994).

6. A. Damasio, T. Grabowski, A. Bechara, H. Damasio, Laura L. B. Ponto, J. Parvizi, and Richard D. Hichwa, "Subcortical and Cortical Brain Activity During the Feeling of Self-generated Emotions," *Nature Neuroscience* 3 (2000), 1049–56.

7. A. Damasio, "Fundamental Feelings," *Nature* 413 (2001), 781; A. Damasio, *Looking for Spinoza* (New York: Harcourt Brace, 2003).

8. See A. D. Craig, "How Do You Feel—Now? The Anterior Insula and Human Awareness," *Nature Reviews Neuroscience* 10 (2009), 59–70. Craig argues that the insular cortex provides the substrate for feeling states, bodily as well as emotional, then goes on to suggest that the very awareness of these states originates in the insula. In direct conflict with Craig's hypothesis is the evidence I adduced in Chapters 3 and 4, on the blatant persistence of feelings and consciousness following insula damage, and on the likely presence of feelings in decorticated children.

9. D. Rudrauf, J. P. Lachaux, A. Damasio, S. Baillet, L. Hugueville, J. Martinerie, H. Damasio, and B. Renault, "Enter Feelings: Somatosensory Responses Following Early Stages of Visual Induction of Emotion," *International Journal of Psychophysiology* 72, no. 1 (2009), 13–23; D. Rudrauf, O. David, J. P. Lachaux, C. Kovach, J. Martinerie, B. Renault, and A. Damasio, "Rapid Interactions Between the Ventral Visual Stream and Emotion-Related Structures Rely on a Two-Pathway Architecture," *Journal of Neuroscience* 28, no. 11 (2008), 2793–803.

10. The original expression is "Quem vê caras não vê corações."

11. A. Damasio, "Neuroscience and Ethics: Intersections," *American Journal of Bioethics* 7, no. 1 (2007), 3–7.

12. M. H. Immordino-Yang, A. McColl, H. Damasio, and A. Damasio, "Neural Correlates of Admiration and Compassion," *Proceedings of the National Academy of Sciences* 106, no. 19 (2009), 8021–26.

13. J. Haidt, "The Emotional Dog and Its Rational Tail: A Social Intuitionist Approach to Moral Judgment," *Psychological Review* 108 (2001), 814–34; Christopher Oveis, Adam B. Cohen, June Gruber, Michelle N. Shiota, Jonathan Haidt, and Dacher Keltner, "Resting Respiratory Sinus Arrhythmia Is Associated with Tonic Positive Emotionality," *Emotion* 9, no. 2 (April 2009), 265–70.

6 / An Architecture for Memory

1. Eric R. Kandel, James H. Schwartz, and Thomas M. Jessel, *Principles of Neural Science,* 4th ed. (New York: McGraw-Hill, 2000); and E. Kandel, *In Search of Memory: The Emergence of a New Science of Mind* (New York: W. W. Norton, 2006).

2. A. R. Damasio, H. Damasio, D. Tranel, and J. P. Brandt, "Neural Regionalization of Knowledge Access: Preliminary Evidence," *Symposia on Quantitative Biology* 55 (1990), 1039–47; A. Damasio, D. Tranel, and H. Damasio, "Face Agnosia and the Neural Substrates of Memory," *Annual Review of Neuroscience* 13 (1990), 89–109.

3. Stephen M. Kosslyn, *Image and Mind* (Cambridge, Mass.: Harvard University Press, 1980).

4. A. R. Damasio, "Time-locked Multiregional Retroactivation: A Systems-level Proposal for the Neural Substrates of Recall and Recognition," *Cognition* 33 (1989), 25–62. The CDZ model has been incorporated in cognitive theories. See, for example,

L. W. Barsalou, "Grounded Cognition," *Annual Review of Psychology* 59 (2008), 617–45, and W. K. Simmons and L. W. Barsalou, "The Similarity-in-Topography Principle: Reconciling Theories of Conceptual Deficits," *Cognitive Neuropsychology* 20 (2003), 451–86.

5. K. S. Rockland and D. N. Pandya, "Laminar Origins and Terminations of Cortical Connections of the Occipital Lobe in the Rhesus Monkey," *Brain Research* 179 (1979), 3–20; G. W. Van Hoesen, "The Parahippocampal Gyrus: New Observations Regarding Its Cortical Connections in the Monkey," *Trends in Neuroscience* 5 (1982), 345–50.

6. Patric Hagmann, Leila Cammoun, Xavier Gigandet, Reto Meuli, Christopher J. Honey, Van J. Wedeen, and Olaf Sporns, "Mapping the Structural Core of Human Cerebral Cortex," *PLoS Biology* 6, no. 7 (2008), e159. doi:10.1371/journal.pbio.0060159.

7. Some convergence zones bind signals relative to entity categories (e.g., the color and shape of a tool) and are placed in association cortices located immediately beyond (downstream from) the cortices whose activity defines featural representations. In humans, in the case of a visual entity, this would include cortices in areas 37 and 39, downstream from the early cortical maps. Their level in the anatomical hierarchy is relatively low. Other CDZs bind signals relative to more complex combinations, for instance, the definition of certain classes of object by binding signals relative to shape, color, sound, temperature, and smell. These CDZs are placed at a higher level in the corticocortical hierarchy (e.g., within anterior sectors of 37, 39, 22, and 20). They stand for combinations of entities or features of varied entities rather than single entities or single features. The CDZs capable of binding entities into events are located at the top of the hierarchical streams, in the most anterior temporal and frontal regions.

8. Kaspar Meyer and Antonio Damasio, "Convergence and Divergence in a Neural Architecture for Recognition and Memory," *Trends in Neurosciences* 32, no. 7 (2009), 376–82.

9. G. A. Calvert, E. T. Bullmore, M. J. Brammer, R. Campbell, S. C. R. Williams, P. K. McGuire, P. W. R. Woodruff, S. D. Iversen, and A. S. David, "Activation of Auditory Cortex During Silent Lip Reading," *Science* 276 (1997), 593–96.

10. M. Kiefer, E. J. Sim, B. Herrnberger, J. Grothe, and K. Hoenig, "The Sound of Concepts: Four Markers for a Link Between Auditory and Conceptual Brain Systems," *Journal of Neuroscience* 28 (2008), 12224–30; J. González, A. Barros-Loscertales, F. Pulvermüller, V. Meseguer, A. Sanjuán, V. Belloch, and C. Ávila, "Reading Cinnamon Activates Olfactory Brain Regions," *NeuroImage* 32 (2006), 906–12; M. C. Hagen, O. Franzen, F. McGlone, G. Essick, C. Dancer, and J. V. Pardo, "Tactile Motion Activates the Human Middle Temporal/V5 (MT/V5) Complex," *European Journal of Neuroscience* 16 (2002), 957–64; K. Sathian, A. Zangaladze, J. M. Hoffman, and S. T. Grafton, "Feeling with the Mind's Eye," *Neuroreport* 8 (1997), 3877–81; A. Zangaladze, C. M. Epstein, S. T. Grafton, and K. Sathian, "Involvement of Visual Cortex in Tactile Discrimination of Orientation," *Nature* 401 (1999), 587–90; Y.-D. Zhou and J. M. Fuster,

"Neuronal Activity of Somatosensory Cortex in a Cross-modal (Visuo-haptic) Memory Task," *Experiments in Brain Research* 116 (1997), 551–55; Y.-D. Zhou and J. M. Fuster, "Visuo-tactile Cross-modal Associations in Cortical Somatosensory Cells," *Proceedings of the National Academy of Sciences* 97 (2000), 9777–82.

11. S. M. Kosslyn, G. Ganis, and W. L. Thompson, "Neural Foundations of Imagery," *Nature Reviews Neuroscience* 2 (2001), 635–42; Z. Pylyshyn, "Return of the Mental Image: Are There Really Pictures in the Brain?" *Trends in Cognitive Science* 7 (2003), 113–18.

12. S. M. Kosslyn, W. L. Thompson, I. J. Kim, and N. M. Alpert, "Topographical Representations of Mental Images in Primary Visual Cortex," *Nature* 378 (1995), 496–98; S. D. Slotnick, W. L. Thompson, and S. M. Kosslyn, "Visual Mental Imagery Induces Retinotopically Organized Activation of Early Visual Areas," *Cerebral Cortex* 15 (2005), 1570–83; S. M. Kosslyn, A. Pascual-Leone, O. Felician, S. Camposano, J. P. Keenan, W. L. Thompson, G. Ganis, K. E. Sukel, and N. M. Alpert, "The Role of Area 17 in Visual Imagery: Convergent Evidence from PET and rTMS," *Science* 284 (1999), 167–70; M. Lotze, and U. Halsband, "Motor Imagery," *Journal of Physiology* 99 (2006), 386–95; K. M. O'Craven and N. Kanwisher, "Mental Imagery of Faces and Places Activates Corresponding Stimulus-specific Brain Regions," *Journal of Cognitive Neuroscience* 12 (2000), 1013–23; M. J. Farah, "Is Visual Imagery Really Visual? Overlooked Evidence from Neuropsychology," *Psychological Review* 95 (1988), 307–17.

13. V. Gallese, L. Fadiga, L. Fogassi, and G. Rizzolatti, "Action Recognition in the Premotor Cortex," *Brain* 119 (1996), 593–609; G. Rizzolatti and L. Craighero, "The Mirror-Neuron System," *Annual Review of Neuroscience* 27 (2004), 169–92.

14. A. Damasio and K. Meyer, "Behind the Looking-Glass," *Nature* 454 (2008), 167–68.

15. A large number of studies from the wide-ranging mirror neuron literature are compatible with the CDZ model: E. Kohler, C. Keysers, M. A. Umiltà, L. Fogassi, V. Gallese, and G. Rizzolatti, "Hearing Sounds, Understanding Actions: Action Representation in Mirror Neurons," *Science* 297 (2002), 846–48; C. Keysers, E. Kohler, M. A. Umiltà, L. Nanetti, L. Fogassi, and V. Gallese, "Audiovisual Mirror Neurons and Action Recognition," *Experiments in Brain Research* 153 (2003), 628–36; V. Raos, M. N. Evangeliou, and H. E. Savaki, "Mental Simulation of Action in the Service of Action Perception," *Journal of Neuroscience* 27 (2007), 12675–83; D. Tkach, J. Reimer, and N. G. Hatsopoulos, "Congruent Activity During Action and Action Observation in Motor Cortex," *Journal of Neuroscience* 27 (2007), 13241–50; S.-J. Blakemore, D. Bristow, G. Bird, C. Frith, and J. Ward, " Somatosensory Activations During the Observation of Touch and a Case of Vision-Touch Synaesthesia," *Brain* 128 (2005), 1571–83; A. Lahav, E. Saltzman, and G. Schlaug, "Action Representation of Sound: Audiomotor Recognition Network While Listening to Newly Acquired Actions," *Journal of Neuroscience* 27 (2007), 308–314; G. Buccino, F. Binkofski, G. R. Fink, L. Fadiga, L. Fogassi, V. Gallese, R. J. Seitz, K. Zilles, G. Rizzolatti, and H.-J. Freund, "Action Observation Activates Premotor and Parietal Areas in a Somatotopic Manner: An fMRI Study,"

European Journal of Neuroscience 13 (2001), 400–04; M. Iacoboni, L. M. Koski, M. Brass, H. Bekkering, R. P. Woods, M.-C. Dubeau, J. C. Mazziotta, and G. Rizzolatti, "Reafferent Copies of Imitated Actions in the Right Superior Temporal Cortex," *Proceedings of the National Academy of Sciences* 98 (2001), 13995–99; V. Gazzola, L. Aziz-Zadeh, and C. Keysers, "Empathy and the Somatotopic Auditory Mirror System in Humans," *Current Biology* 16 (2006), 1824–29; C. Catmur, V. Walsh, and C. Heyes, "Sensorimotor Learning Configures the Human Mirror System," *Current Biology* 17 (2007), 1527–31; C. Catmur, H. Gillmeister, G. Bird, R. Liepelt, M. Brass, and C. Heyes, "Through the Looking Glass: Counter-Mirror Activation Following Incompatible Sensorimotor Learning," *European Journal of Neuroscience* 28 (2008), 1208–15.

16. G. Kreiman, C. Koch, and I. Fried, "Imagery Neurons in the Human Brain," *Nature* 408 (2000), 357–61.

7 / Consciousness Observed

1. Harold Bloom, *The Western Canon* (New York: Harcourt Brace, 1994); Harold Bloom, *Shakespeare: The Invention of the Human* (New York: Riverhead, 1998); James Wood, *How Fiction Works* (New York: Farrar, Straus and Giroux, 2008).

2. For recent reviews of the basic neuroscience of consciousness, I recommend *The Neurology of Consciousness,* ed. Steven Laureys and Giulio Tononi (London: Elsevier, 2008). For reviews on the clinical aspects of consciousness, I recommend Jerome B. Posner, Clifford B. Saper, Nicholas D. Schiff, and Fred Plum, *Plum and Posner's Diagnosis of Stupor and Coma,* (2007), cited earlier. Also see Todd E. Feinberg, *Altered Egos: How the Brain Creates the Self* (New York: Oxford University Press, 2001), for a recent review of the relevant clinical literature; and A. R. Damasio, "Consciousness and Its Disorders," in *Diseases of the Nervous System: Clinical Neuroscience and Therapeutic Principles,* ed. Arthur K. Asbury, G. McKhann, I. McDonald, P. J. Goadsby, and J. McArthur, 3rd ed. (New York: Cambridge University Press, 2002), 2, 289–301.

3. Adrian Owen, "Detecting Awareness in the Vegetative State," *Science* 313 (2006), 1402.

4. Adrian Owen and Steven Laureys, "Willful Modulation of Brain Activity in Disorders of Consciousness," *New England Journal of Medicine* 362 (2010), 579–89.

5. Antonio Damasio, *The Feeling of What Happens: Body and Emotion in the Making of Consciousness* (New York: Harcourt, Brace, 1999).

6. Antonio Damasio, "The Somatic Marker Hypothesis and the Possible Functions of the Prefrontal Cortex," *Philosophical Transactions of the Royal Society B: Biological Sciences* 351 (1996), 1413–20.

7. Sigmund Freud, "Some Elementary Lessons in Psychoanalysis," *International Journal of Psycho-Analysis* 21 (1940).

8. Kraft-Ebbing, *Psychopathia Sexualis* (Stuttgart: Ferdinand Enke, 1886).

9. For thoughtful considerations on mind and consciousness during sleep and dreaming, I recommend Allan Hobson's *Dreaming: An Introduction to the Science of Sleep* (New York: Oxford University Press, 2002), and Rodolfo Llinás, *I of the Vortex: From Neurons to Self* (Cambridge, Mass.: MIT Press, 2002).

8 / Building a Conscious Mind

1. Bernard Baars is a good example of this approach, which has been used with advantage by Changeux and Dehaene. See S. Dehaene, M. Kerszberg, and J.-P. Changeux, "A Neuronal Model of a Global Workspace in Effortful Cognitive Tasks," *Proceedings of the National Academy of Sciences* 95, no. 24 (1998), 14529–34. Edelman and Tononi have also approached consciousness from this perspective. See Gerald M. Edelman and Giulio Tononi, *A Universe of Consciousness: How Matter Becomes Imagination* (New York: Basic Books, 2000). Likewise, the work of Crick and Koch focuses on the mind aspects of consciousness and explicitly acknowledges that the self is not part of the agenda. See F. Crick and C. Koch, "A Framework for Consciousness," *Nature Neuroscience* 6, no. 2 (2003), 119–26.

2. I am thinking of these extremely important studies: G. Moruzzi and H. W. Magoun, "Brain Stem Reticular Formation and Activation of the EEG," *Electroencephalography and Clinical Neurophysiology* 1 (1949): 455–73; and W. Penfield and H. H. Jasper, *Epilepsy and the Functional Anatomy of the Human Brain* (New York: Little, Brown, 1954).

3. As stated in note 17 of Chapter 1, Panksepp also gives emphasis to the notion of early feelings, without which the process of consciousness cannot proceed. The detailed mechanism is not the same, but I believe the essence of the idea is. More often than not, treatments of feeling assume that they arise from interactions with the world (as in James's "feelings of knowing" or my "feeling of what happens") or as a result of emotions. But primordial feelings *precede* those situations, and presumably Panksepp's early feelings do too.

4. L. W. Swanson, "The Hypothalamus," in *Handbook of Chemical Neuroanatomy*, vol. 5, *Integrated systems of the CNS,* ed. A. Björklund, T. Hökfelt, and L. W. Swanson (Amsterdam: Elsevier, 1987).

5. J. Parvizi and A. Damasio, *Cognition*. See extensive discussion in Antonio Damasio, *The Feeling of What Happens: Body and Emotion in the Making of Consciousness* (New York: Harcourt, Brace, 1999).

6. Bernard J. Baars, "Global Workspace Theory of Consciousness: Toward a Cognitive Neuroscience of Human Experience," *Progress in Brain Research* 150 (2005), 45–53; D. L. Sheinberg and N. K. Logothetis, "The Role of Temporal Cortical Areas in Perceptual Organization," *Proceedings of the National Academy of Sciences* 94, no. 7 (1997), 3408–13; S. Dehaene, L. Naccache, L. Cohen, et al., "Cerebral Mechanisms of Word Masking and Unconscious Repetition Priming," *Nature Neuroscience* 4, no. 7 (2001), 752–58.

7. As noted in Chapter 5, the contributions of A. D. Craig regarding the spinal cord and cortical aspects of the system are especially noteworthy: A. D. Craig, "How Do You Feel? Interoception: The Sense of the Physiological Condition of the Body," *Nature Reviews Neuroscience* 3 (2002), 655–66.

8. K. Meyer, "How Does the Brain Localize the Self?" *Science E-letters* (2008), available at www.sciencemag.org/cgi/eletters/317/5841/1096#10767. See also B. Lenggenhager, T. Tadi, T. Metzinger, and O. Blanke, "Video Ergo Sum: Manipulating Bodily Self-Consciousness," *Science* 317 (2007), 1096; and H. H. Ehrsson, "The Experimental Induction of Out-of-Body Experiences," *Science* 317 (2007), 1048.

9. Michael Gazzaniga, *The Mind's Past* (Berkeley: University of California Press, 1998).

10. My interest in the superior colliculi goes back to the mid-1980s. Someone who was even more intrigued by the colliculi was Bernard Strehler, with whom I discussed the issue on several occasions. More recently Bjorn Merker has presented a compelling picture of this structure as more than a mere assistant to vision. Bernard M. Strehler, "Where Is the Self? A Neuroanatomical Theory of Consciousness," *Synapse* 7 (1991), 44–91; Bjorn Merker, "Consciousness Without a Cerebral Cortex," *Behavioral and Brain Sciences* 30 (2007), 63–81. In his discussion of the importance of the periaqueductal gray, Jaak Panksepp has also called attention to the colliculi.

11. The building of sensory perspective would result from combining newly obtained images of the pelicans with activity in the sensory portals engaged by the organism-object interaction. Sensory portal activity would be linked with images of the object by synchronizing the activities related to each set of images. Time, not space, would be the critical link. The sense of agency and of owning one's mind would be derived by a comparable mechanism, linking in time the activities that pertain to new object images with those that define changes in the protoself at the level of interoceptive maps, sensory portals, and musculoskeletal representations. The degree of cohesion with which these components would be held would depend on timing.

9 / The Autobiographical Self

1. C. Koch and F. Crick, "What Is the Function of the Claustrum?" *Philosophical Transactions of the Royal Society B: Biological Sciences* 360, no. 1458 (June 29, 2005), 1271–79.

2. R. J. Maddock, "The Retrosplenial Cortex and Emotion: New Insights from Functional Neuroimaging of the Human Brain," *Trends in Neurosciences* 22 (1999), 310–16; R. Morris, G. Paxinos, and M. Petrides, "Architectonic Analysis of the Human Retrosplenial Cortex," *Journal of Comparative Neurology* 421 (2000), 14–28; for a review, see A. E. Cavanna and M. R. Trimble, "The Precuneus: A Review of Its Functional Anatomy and Behavioural Correlates," *Brain* 129 (2006), 564–83.

3. J. Parvizi, G. W. Van Hoesen, J. Buckwalter, and A. R. Damasio, "Neural Connections of the Posteromedial Cortex in the Macaque," *Proceedings of the National Academy of Sciences* 103 (2006), 1563–68.

4. Patric Hagmann, Leila Cammoun, Xavier Gigandet, Reto Meuli, Christopher J. Honey, Van J. Wedeen, and Olaf Sporns, "Mapping the Structural Core of Human Cerebral Cortex," *PLoS Biology 6, e159. doi:10.1371/journal.pbio.0060159.*

5. Pierre Fiset, Tomás Paus, Thierry Daloze, Gilles Plourde, Pascal Meuret, Vincent Bohnomme, Nadine Hajj-Ali, Steven B. Backman, and Alan C. Evans, "Brain Mechanisms of Propofol-induced Loss of Consciousness in Humans: A Positron Emission Tomographic Study," *Journal of Neuroscience* 19 (2009), 5506–13; M. T. Alkire and J. Miller, "General Anesthesia and the Neural Correlates of Consciousness," *Progress in Brain Research* 150 (2005), 229–44. Propofol's success at turning off consciousness is not far away from turning off life altogether—one reason the monitoring of this drug's effects must be so careful. Michael Jackson appears to have died from an overdose of propofol or possibly from an unfortunate combination of propofol with other brain-active medications.

6. Pierre Maquet, Christian Degueldre, Guy Delfiore, Joël Aerts, Jean-Marie Péters, André Luxen, and Georges Franck, "Functional Neuroanatomy of Human Slow Wave Sleep," *Journal of Neuroscience* 17 (1997), 2807–12; P. Maquet et al., "Human Cognition During REM Sleep and the Activity Profile Within Frontal and Parietal Cortices: A Reappraisal of Functional Neuroimaging Data," *Progress in Brain Research* 150 (2005), 219–27; M. Massimini et al., "Breakdown of Cortical Effective Connectivity During Sleep," *Science* 309 (2005), 2228–32.

7. D. A. Gusnard and M. E. Raichle, "Searching for a Baseline: Functional Imaging and the Resting Human Brain," *Nature Reviews Neuroscience* 2 (2001), 685–94.

8. Antonio R. Damasio, Thomas J. Grabowski, Antoine Bechara, Hanna Damasio, Laura L.B. Ponto, Josef Parvizi, and Richard D. Hichwa, "Subcortical and Cortical Brain Activity During the Feeling of Self-generated Emotions," *Nature Neuroscience* 3 (2000), 1049–56.

9. R. L. Buckner and Daniel C. Carroll, "Self-projection and the Brain," *Trends in Cognitive Sciences* 11, no. 2 (2006), 49–57; R. L. Buckner, J. R. Andrews-Hanna, and D. L. Schacter, "The Brain's Default Network: Anatomy, Function, and Relevance to Disease," *Annals of the New York Academy of Sciences* 1124 (2008), 1–38; M. H. Immordino-Yang, A. McColl, H. Damasio, et al., "Neural Correlates of Admiration and Compassion," *Proceedings of the National Academy of Sciences* 106, no. 19 (2009), 8021–26; R. L. Buckner et al., "Cortical Hubs Revealed by Intrinsic Functional Connectivity: Mapping, Assessment of Stability, and Relation to Alzheimer's Disease," *Journal of Neuroscience* 29 (2009), 1860–73.

10. M. E. Raichle and M. A. Mintun, "Brain Work and Brain Imaging," *Annual Review of Neuroscience* 29 (2006), 449–76; M. D. Fox et al., "The Human Brain Is Intrinsically Organized into Dynamic, Anticorrelated Functional Networks," *Proceedings of the National Academy of Sciences* 102 (2005), 9673–78.

11. B. T. Hyman, G. W. Van Hoesen, and A. R. Damasio, "Cell-specific Pathology Isolates the Hippocampal Formation," *Science* 225 (1984), 1168–70; G. W. Van Hoesen,

B. T. Hyman, and A. R. Damasio, "Cellular Disconnection Within the Hippocampal Formation as a Cause of Amnesia in Alzheimer's," *Neurology* 34, no. 3 (1984), 188–89; G. W. Van Hoesen and A. Damasio, "Neural Correlates of Cognitive Impairment in Alzheimer's Disease," in *Handbook of Physiology, Higher Functions of the Brain,* ed. V. Mountcastle and F. Plum (Bethesda, Md.: American Physiological Society, 1987).

12. J. Parvizi, G. W. Van Hoesen, and A. R. Damasio, "Selective Pathological Changes of the Periaqueductal Gray in Alzheimer's Disease," *Annals of Neurology* 48 (2000), 344–53; J. Parvizi, G. W. Van Hoesen, and A. Damasio, "The Selective Vulnerability of Brainstem Nuclei to Alzheimer's Disease," *Annals of Neurology* 49 (2001), 53–66.

13. R. L. Buckner et al., "Molecular, Structural, and Functional Characterization of Alzheimer's Disease: Evidence for a Relationship Between Default Activity, Amyloid, and Memory," *Journal of Neuroscience* 25 (2005), 7709–17; S. Minoshima et al., "Metabolic Reduction in the Posterior Cingulate Cortex in Very Early Alzheimer's Disease," *Annals of Neurology* 42 (1997), 85–94.

14. Curiously, the fact that the PMCs are involved in Alzheimer's disease turns out to be an old but overlooked finding, identified as early as 1976. See A. Brun and L. Gustafson, "Distribution of Cerebral Degeneration in Alzheimer's Disease," *European Archives of Psychiatry and Clinical Neuroscience* 223, no. 1 (1976). Brun and Gustafson had called attention to the notable contrast between the intact anterior cingulate cortex (it is usually spared in Alzheimer's) and the posterior cingulate cortex, where there was abundant pathology. They could not have known then that neurofibillary tangles in the PMCs came later in the course of the disease than the anterior temporal damage; nor did they know what we know today regarding the internal structure of the PMCs and their peculiar wiring diagram. See A. Brun and E. Englund, "Regional Pattern of Degeneration in Alzheimer's Disease: Neuronal Loss and Histopathological Grading," *Histopathology* 5 (1981), 549–64; A. Brun and L. Gustafson, "Limbic Involvement in Presenile Dementia," *Archiv für Psychiatrie und Nervenkrankheiten* 226 (1978), 79–93.

15. G. W. Van Hoesen, B. T. Hyman, and A. R. Damasio, "Entorhinal Cortex Pathology in Alzheimer's Disease," *Hippocampus* 1 (1991), 1–8.

16. Randy Buckner and his colleagues have described this possibility as the "metabolism hypothesis." Buckner's group has also presented compelling functional neuroimaging evidence to the effect that the PMCs show prominent decreases of glucose metabolism as Alzheimer's disease advances.

17. J. D. Bauby, *Le Scaphandre et le papillon* (Paris: Éditions Robert Laffont, 1997).

18. S. Laureys et al., "Differences in Brain Metabolism Between Patients in Coma, Vegetative State, Minimally Conscious State and Locked-in Syndrome," *European Journal of Neurology* 10 (suppl 1.) (2003), 224–25; and S. Laureys, "The Neural Correlate of (Un)awareness: Lessons from the Vegetative State," *Trends in Cognitive Sciences* 9 (2005), 556–59.

19. S. Laureys, M. Boly, and P. Maquet, "Tracking the Recovery of Consciousness from Coma," *Journal of Clinical Investigation* 116 (2006), 1823–25.

20. A. D. Craig, "How Do You Feel—Now? The Anterior Insula and Human Awareness," *Nature Reviews Neuroscience* 10 (2009), 59–70.

10 / Putting It Together

1. Jerome B. Posner, Clifford B. Saper, Nicholas D. Schiff, and Fred Plum, *Plum and Posner's Diagnosis of Stupor and Coma* (New York: Oxford University Press, 2007).

2. J. Parvizi and A. R. Damasio, "Neuroanatomical Correlates of Brainstem Coma," *Brain* 126 (2003), 1524–36.

3. G. Moruzzi and H. W. Magoun, "Brain Stem Reticular Formation and Activation of the EEG," *Electroencephalography and Clinical Neurophysiology* 1 (1949), 455–73; J. Olszewski, "Cytoarchitecture of the Human Reticular Formation," in *Brain Mechanisms and Consciousness,* ed. J. F. Delafresnaye et al. (Springfield, Ill.: Charles C. Thomas, 1954); A. Brodal, *The Reticular Formation of the Brain Stem: Anatomical Aspects and Functional Correlations* (Edinburgh: William Ramsay Henderson Trust, 1959); A. N. Butler and W. Hodos, "The Reticular Formation," in *Comparative Vertebrate Neuroanatomy: Evolution and Adaptation,* ed. Ann B. Butler and William Hodos (New York: Wiley-Liss, 1996); and W. Blessing, "Inadequate Frameworks for Understanding Bodily Homeostasis," *Trends in Neurosciences* 20 (1997), 235–39.

4. J. Parvizi and A. Damasio, "Consciousness and the Brainstem," *Cognition* 49 (2001), 135–59.

5. E. G. Jones, *The Thalamus,* 2nd ed. (New York: Cambridge University Press, 2007); Rodolfo Llinás, *I of the Vortex: From Neurons to Self* (Cambridge, Mass.: MIT Press, 2002); M. Steriade and M. Deschenes, "The Thalamus as a Neuronal Oscillator," *Brain Research* 320 (1984), 1–63; M. Steriade, "Arousal: Revisiting the Reticular Activating System," *Science* 272 (1992), 225–26.

6. A comprehensive review of the fundamentals of cerebral cortex anatomy and physiology is available in a major collection of articles: E. G. Jones, A. Peters, and John H. Morrison, eds., *Cerebral Cortex* (New York: Springer, 1999).

7. Several contemporary philosophers who have dealt with the mind-body problem have addressed the issues of qualia in one way or another. The following work has been of special value to me: John R. Searle, *The Mystery of Consciousness* (New York: New York Review Books, 1990); Patricia Churchland, *Neurophilosophy: Toward a Unified Science of the Mind-Brain* (Cambridge, Mass.: MIT Press, 1989); R. McCauley, ed., *The Churchlands and their Critics* (New York: Wiley-Blackwell, 1996); D. Dennet, *Consciousness Explained* (New York: Little, Brown, 1992); Simon Blackburn, *Think: A Compelling Introduction to Philosophy* (Oxford: Oxford University Press, 1999); Ned Block, ed., *The Nature of Consciousness: Philosophical Debates* (Cambridge, Mass.: MIT Press, 1997); Owen Flanagan, *The Really Hard Problem: Meaning in a Material World* (Cambridge, Mass.: MIT Press, 2007); T. Metzinger, *Being No One: The Self-Model Theory of Subjectivity* (Cambridge, Mass.: MIT Press, 2003); David Chalmers, *The Conscious*

Mind: In Search of a Fundamental Theory (Oxford: Oxford University Press, 1996); Galen Strawson, "The Self," *Journal of Consciousness Studies* 4 (1997), 405–28; and Thomas Nagel, "What Is it Like to Be a Bat?" *Philosophical Review* (1974), 435–50.

8. Llinás, *Vortex*.

9. N. D. Cook, "The Neuron-level Phenomena Underlying Cognition and Consciousness: Synaptic Activity and the Action Potential," *Neuroscience* 153 (2008), 556–70.

10. R. Penrose, *The Emperor's New Mind: Concerning Computers, Minds, and the Laws of Physics* (Oxford: Oxford University Press, 1989); S. Hameroff, "Quantum Computation in Brain Microtubules? The Penrose-Hameroff 'Orch OR' Model of Consciousness," *Philosophical Transactions of the Royal Society A: Mathematical, Physical and Engineering Sciences* 356 (1998), 1869–96.

11. D. T. Kemp, "Stimulated Acoustic Emissions from Within the Human Auditory System," *Journal of the Acoustical Society of America* 64, no. 5 (1978), 1386–91.

12. One of the puzzles of the Qualia II problem centers on the assumption that neurons that are similar among themselves would not produce neural states that are qualitatively different. The argument, however, is fallacious. The general operation of neurons is formally similar, to be sure, but the neurons of distinct sensory systems are vastly different in kind. They emerged at different ages in evolution, and the profile of their activities is likely to be distinct as well. Neurons involved in body sensing might well have special characteristics that would play a role in the generation of feelings. Moreover, the patterns of their interactivity with other regions, even within the same sensory cortical complex, vary greatly.

We are barely beginning to understand the microcircuitry of our peripheral sensory devices, and we know even less about the microcircuitry of the subcortical stations and cortical areas that map out of the initial data generated at the sensory devices themselves. We still know very little about the connectivity among those separate stations, especially about the connectivity that happens in reverse, from the brain toward the periphery. Why, for example, does the primary visual cortex (V_1 or area 17) send more projections down to the lateral geniculate nucleus than the nucleus itself sends to the cortex? This is quite strange. The brain is in the business of collecting signals *from* the outside world and bringing them into its structures. These "downward and outward" pathways must be accomplishing something useful, or they would have been weeded out in evolution. They remain unexplained. Feedback correction is the standard account for "back" projections, but why should signal correction be the whole explanation? Within the cerebral cortex itself, I believe that back projections work as "retroactivators," as suggested in the convergence-divergence framework. For example, besides all the signals coming from the eyeball and surround, does the retina also send to the brain signals other than visual, for example, somatosensory information? A sizable part of the answer to why seeing red is different from hearing a cello or smelling cheese may come from such additional understanding.

11 / Living with Consciousness

1. A large body of literature speaks to these findings, beginning with H. H. Kornhuber and L. Deecke, "Hirnpotentialänderungen bei Willkürbewegungen und passiven Bewegungen des Menschen: Bereitschaftspotential und reafferente Potentiale," *Pflugers Archiv für Gesamte Psychologie* 284 (1965), 1–17; B. Libet, C. A. Gleason, E. W. Wright, and D. K. Pearl, "Time of Conscious Intention to Act in Relation to Onset of Cerebral Activity (Readiness-potential)," *Brain* 106 (1983), 623–42; B. Libet, "Unconscious Cerebral Initiative and the Role of Conscious Will in Voluntary Action," *Behavior and Brain Sciences* 8 (1985), 529–66.

Other important contributors to the literature on these issues include: D. M. Wegner, *The Illusion of Conscious Will* (Cambridge, Mass.: MIT Press, 2002); P. Haggard and M. Eimer, "On the Relationship Between Brain Potentials and the Awareness of Voluntary Movements," *Experimental Brain Research* 126 (1999), 128–133; C. D. Frith, K. Friston, P. F. Liddle, and R. S. J. Frackowiak, "Willed Action and the Prefrontal Cortex in Man: A Study with PET," *Proceedings of the Royal Society of London, Series B* 244 (1991), 241–46; R. E. Passingham, J. B. Rowe, and K. Sakai, "Prefrontal Cortex and Attention to Action," in *Attention in Action,* ed. G. Humphreys and M. Riddoch (New York: Psychology Press, 2005).

2. A very well-argued review of this problem is C. Suhler and P. Churchland, "Control: Conscious and Otherwise," *Trends in Cognitive Sciences* 13 (2009), 341–47. See also J. A. Bargh, M. Chen, and L. Burrows, "Automaticity of Social Behavior: Direct Effects of Trait Construct and Stereotype Activation on Action," *Journal of Personality and Social Psychology* 71 (1996), 230–44; R. F. Baumeister et al., "Self-regulation and the Executive Function: The Self as Controlling Agent," *Social Psychology: Handbook of Basic Principles,* 2nd ed., ed. A. Kruglanski and E. Higgins (New York: Guilford Press, 2007); R. Poldrack et al., "The Neural Correlates of Motor Skill Automaticity," *Journal of Neuroscience* 25 (2005), 5356–64.

3. S. Gallagher, "Where's the Action? Epiphenomenalism and the Problem of Free Will," in *Does Consciousness Cause Behavior?* ed. Susan Pockett, William P. Banks, and Shaun Gallagher (Cambridge, Mass.: MIT Press, 2009).

4. Ap Dijksterhuis, "On Making the Right Choice: The Deliberation-without-Attention Effect," *Science* 311 (2006), 1005.

5. A. Bechara, A. R. Damasio, H. Damasio, and S. W. Anderson, "Insensitivity to Future Consequences Following Damage to Prefrontal Cortex," *Cognition* 50 (1994), 7–15; A. Bechara, H. Damasio, D. Tranel, and A. R. Damasio, "Deciding Advantageously Before Knowing the Advantageous Strategy," *Science* 275 (1997), 1293–94.

6. A recent set of experiments from Alan Cowey's laboratory confirms, using a waging paradigm, that the choice of winning strategy in our gambling experiment is processed nonconsciously. N. Persaud, P. McLeod, and A. Cowey, "Post-decision Wagering Objectively Measures Awareness," *Nature Neuroscience* 10, no. 2 (2007), 257–61.

7. D. Kahneman, "Maps of Bounded Rationality: Psychology for Behavioral Economists," *American Economic Review* 93 (2003), 1449–75; D. Kahneman and S. Frederick, "Frames and Brains: Elicitation and Control of Response Tendencies," *Trends in Cognitive Science* 11 (2007), 45–46; Jason Zweig, *Your Money and Your Brain: How the New Science of Neuroeconomics Can Help Make You Rich* (New York: Simon and Schuster, 2007); and J. Lehrer, *How We Decide* (New York: Houghton Mifflin, 2009).

8. Elizabeth A. Phelps, Christopher J. Cannistraci, and William A. Cunningham, "Intact Performance on an Indirect Measure of Race Bias Following Amygdala Damage," *Neuropsychologia* 41, no. 2 (2003), 203–08; N. N. Oosterhof and A. Todorov, "The Functional Basis of Face Evaluation," *Proceedings of the National Academy of Sciences* 105 (2008), 11087–92. Evidence for nonconscious biases is also well covered in intelligent popular writing.

9. Wegner, *Illusion*.

10. T. H. Huxley, "On the Hypothesis That Animals Are Automata, and Its History," *Fortnightly Review* 16 (1874), 555–80; reprinted in *Methods and Results: Essays by Thomas H. Huxley* (New York: D. Appleton, 1898).

11. The McArthur Foundation has launched an ambitious project on neuroscience and the law, based on a large consortium of institutions. Led by Michael Gazzaniga, it aims at surveying, debating, and investigating some of these issues in light of contemporary neuroscience.

12. Pertinent work from our group includes: S. W. Anderson, A. Bechara, H. Damasio, D. Tranel, and A. R. Damasio, "Impairment of Social and Moral Behavior Related to Early Damage in Human Prefrontal Cortex," *Nature Neuroscience* 2, no. 11 (1999), 1032–37; M. Koenigs, L. Young, R. Adolphs, D. Tranel, M. Hauser, F. Cushman, and A. Damasio, "Damage to the Prefrontal Cortex Increases Utilitarian Moral Judgments," *Nature* 446 (2007), 908–11; A. Damasio, "Neuroscience and Ethics: Intersections," *American Journal of Bioethics* 7 (2007), 1, 3–7; L. Young, A. Bechara, D. Tranel, H. Damasio, M. Hauser, and A. Damasio, "Damage to Ventromedial Prefrontal Cortex Impairs Judgment of Harmful Intent," *Neuron* 65, no. 6 (2010), 845–51.

13. Julian Jaynes, *The Origin of Consciousness in the Breakdown of the Bicameral Mind* (New York: Houghton Mifflin, 1976).

14. Two recent and very different volumes present an intelligent view of the origins, historical development, and biological underpinnings of religious thinking: Richard Wright, *The Evolution of God* (New York: Little, Brown, 2009); and Nicholas Wade, *The Faith Instinct* (New York: Penguin Press, 2009).

15. W. H. Durham, *Co-evolution: Genes, Culture and Human Diversity* (Palo Alto, Calif.: Stanford University Press, 1991); C. Holden and R. Mace, "Phylogenetic Analysis of the Evolution of Lactose Digestion in Adults," *Human Biology* 69 (1997), 605–28; Kevin N. Laland, John Odling-Smee, and Sean Myles, "How Culture Shaped the Human Genome: Bringing Genetics and the Human Sciences Together," *Nature Reviews Genetics* 11 (2010), 137–48.

16. The biologist E. O. Wilson first called attention to the evolutionary significance of these features. Dennis Dutton provides a comprehensive list of such critical features in *The Art Instinct: Beauty, Pleasure, and Human Evolution* (New York: Bloomsbury Press, 2009). He too presents a biological perspective on the origins of the arts, although his emphasis is on cognitive aspects and mine is on homeostasis.

17. T. S. Eliot, *The Four Quartets* (New York: Harcourt Books, 1968). These words are from the last three verses of Part I in the "Burnt Norton" section.

ACKNOWLEDGMENTS

Architects will tell you that God made nature and architects made the rest, a good way of reminding us that places and spaces, natural and built by humans, play a major role in who we are and what we do. I began this book on a wintry Paris morning, wrote most of the text over the two subsequent summers in Malibu, and I am writing these lines and reviewing proofs during yet another summer, in East Hampton. Since places do count, my first heartfelt thanks go to ever-festive Paris, never mind the snow and the gray; to Cori and Dick Lowe, for the paradise they have created over the Pacific (with some help from Richard Neutra); and to Courtney Ross and the very different version of paradise she has assembled on the other coast, with her exquisite taste.

The backdrop for a book on science, however, goes well beyond a sense of place. In my case it has most to do with the colleagues and students that I have been fortunate to have at the University of Southern California, both within the Brain and Creativity Institute and the Dornsife Cognitive Neuroscience Imaging Center, as well as in several other USC departments and schools. My thanks, then, to the leadership of USC's College of Letters, Arts and Sciences; to Dana and David Dornsife; and to Lucy Billingsley and Joyce Cammilleri, whose support has been vital to creating our everyday intellectual environment. Equally important thanks are due to the research funding agencies that make our work possible, most especially the National Institute for Neurological Disorders and Stroke, and the Mathers Foundation.

Some colleagues and friends read through the whole manuscript or varied parts of it, made suggestions, and discussed the substance of its ideas. They are: Hanna Damasio, Kaspar Meyer, Charles Rockland, Ralph Greenspan, Caleb Finch, Michael Quick, Manuel Castells, Mary

Helen Immordino-Yang, Jonas Kaplan, Antoine Bechara, Rebecca Rickman, Sidney Harman, Gary W. Van Hoesen, and Bruce Adolphe. An even wider group was kind enough to read the material and give me the benefit of their reactions or suggestions. They are: John Allen, Ursula Bellugi, Michael Carlisle, Patricia Churchland, Maria de Sousa, Helder Filipe, Stephan Heck, Siri Hustvedt, Jane Isay, Jonah Lehrer, Yo-Yo Ma, Kingson Man, Joseph Parvizi, Peter Sacks, Julião Sarmento, Peter Sellars, Daniel Tranel, Koen van Gulik, and Bill Viola. My gratitude to all, for their wisdom, frankness, and generosity. The many omissions and failures that remain are my responsibility, not theirs.

Dan Frank, my editor at Pantheon, is a man of several editorial personalities, at least three of which I am able to diagnose—the philosopher, the scientist, and the novelist. Each of these surfaced as needed to dispense gentle but firm influential advice on the manuscript. I am grateful for his counsel and for the patience with which he waited for my fussy amendments. And I am thankful, as always, to Michael Carlisle, longtime friend, adopted brother, and agent, for his wisdom, intelligence, and loyalty, and to Alexis Hurley, who extends him in the world at large.

I thank Kaspar Meyer for preparing Figures 6.1 and 6.2, and Hanna Damasio for preparing all the other figures as well as allowing me to use, in Chapter 4, ideas and some wording from an article on mind and body we wrote together for *Daedalus* some years ago. Gary W. Van Hoesen was kind enough to give us the photograph on which Figure 9.6 is based.

Cinthya Nuñez prepared the manuscript patiently, proficiently, and with great cheer, over countless revisions; Ryan Essex, Pamela McNeff, and Susan Lynch helped competently with indispensible library research. My thanks for their invaluable efforts.

Ethan Bassoff and Lauren Smythe, at Inkwell Management, lent their sympathetic ears and professional brains to all my questions and requests, as did many in the Knopf/Pantheon publishing team, most especially the ever-smiling and enthusiastic Michiko Clark, Jillian Verrillo, Janet Biehl, Andrew Dorko, and Virginia Tan. My thanks to all for their contributions to the final product.

INDEX

Page numbers in *italic* refer to captions.

abstract images, 70–71

acetylcholine, *193,* 225

action potential, 303

addiction, 272, 282

admiration, 125, 126–29

Adolphs, R., 329*n*12, 341*n*12

agency, 165, 185, *206,* 209, 335*n*11

akinetic mutism, 237

Alkire, M. T., 336*n*5

Alzheimer's disease, 229–33, 337*n*14

amnesia, 237–38

amoeba, 33, 257–58

amygdala
  anatomy, 306, 309, 311
  in as-if body loop mechanism, 102, 103
  in emotional processing, 112
  in fear response, 113
  in production of qualia effects, 255

amyotropic lateral sclerosis, 234–35

anatomy
  as aggregation of systems, 33–34
  engineering metaphors for, 44–45
  see also brain structure and function

Andrews-Hanna, J. R., 336*n*9

A nerve fibers, 96, 97

anesthesia, 225–26

animals
  adaptive behavior without
    consciousness in, 31–32
  consciousness in, 26, 171–72
  levels of self in, 26
  manifestations of social emotions in,
    126
  see also simple life forms

anosognosia, 239

anterior cingulate cortex, in emotion
  and feeling, 118, 127, 326*n*8

Aplysia californica, 32

area postrema, 307–9

arts, 278–79, 288–89, 290, 294–96

Asbury, Arthur K., 333*n*2

ascending reticular activating system,
  245–46

as-if body loop mechanism, 102–4
  in creation of feeling of emotion,
    120–21

asomatognosia, 239–40

association cortices. *See* early sensory
  cortices

attention
  brain activity during and after,
    227–28
  in creation of core self, 203
  definition, 203
  emotion effects on, 110–11

auditory system
  brain structures involved in, 84,
    326–27*n*12
  components of sensory portal in,
    197
  construction of perceptual quality in,
    260–61
  hearing versus listening, 173
  image space for memory recall in, 142
  map making in, 68–69
  see also sensory systems

autobiographical consciousness, 168–69,
  171

Diving Bell and the Butterfly, The, 235
dopaminergic system, 47, *192–93*
Doya, Kenji, 324n8
dreams
  brain activity in, 227
  concept of consciousness and, 158, 178
  construction of maps in, 64
  flow of images in, 71
  Freudian conceptualization of, 177–78
  memory of, 178–79
drives
  emotion and, 109, 111
  genomic unconscious in, 278
  mechanisms for homeostasis and, 55
drug effects, 121, 170, 254–55, 282
dura mater, 97
Durham, W. H., 341n15
Dutton, Dennis, 342n16

early sensory cortices
  anatomy, 310
  components of, 311–12
  functions, 310
  in generation of core self state, 205–6
  image making in, 75
  image spaces composed of, 142–43, 271
  in memory processes, 137, 138, 140, 141, 150
Eckhorn, R., 87
Edelman, Gerald, 321n10, 322n15, n16, 323n1, 325n3, 334n1
efference copy, 102
Einstein, Albert, 319–20n1
electrochemical signaling
  in body-brain interaction, 96, 307–9
  in body mapping, 92–93
  in generation of feeling states, 260
  in mind-making regions of brain, 86–88
  for monitoring of body's interior, 96–97, 259–60
  neuronal, 37–38, 285, 302–3
  timing and synchronization of, 87

electroencephalography, 161, 226
Eliot, T. S., 297, 320n6, 342n17
embarrassment, 125
emotion
  background emotions, 125
  basic automated program of, 123
  biological cost of, 114
  biological value and, 108
  brain-body communication to induce, 96
  brain structures and processes in, 74, 80, 82, 110, 112, 113–14, 167
  in changes in memories over time, 211
  classification of, 122–23
  cognitive processes in, 109, 110–11, 116, 119
  control of, 124–25
  creation of somatic markers, 9, 175
  definition, 109
  early human existence, 291
  emotion-feeling cycle, 110–11
  evolutionary origins of, 44, 123–24, 126
  feelings versus, 109–10, 115
  in image management, 174–75
  as indicator of consciousness, 166, 167
  individual differences in experiencing and responding to, 124
  James's conceptualization of, 114–16
  mechanisms for homeostasis in origins of, 55
  in nonconscious cognitive processing, 276–77, 282
  self-concerned life regulation as basis for, 59–60
  significance of, in brain and mind concepts, 108, 111
  specificity of response in, 112–13
  timing of processing of, 122
  triggering, 110, 111–12, 255
  universality of, 123
  see also emotional expressiveness; feelings of emotion; social emotions

Antonio Damasio is David Dornsife Professor of Neuroscience, Psychology and Neurology, and director of the Brain and Creativity Institute, at the University of Southern California. Damasio's books include *Descartes' Error: Emotion, Reason, and the Human Brain; The Feeling of What Happens: Body and Emotion in the Making of Consciousness* (named one of the ten best books of the year by the *New York Times Book Review*); and *Looking for Spinoza: Joy, Sorrow, and the Feeling Brain,* which have been translated into more than thirty languages and taught worldwide. He is the recipient of numerous honors and awards including the Pessoa, Signoret, and Cozzarelli prizes (shared with his wife, Hanna), and the Prince of Asturias Award for Technical and Scientific Research. He is a fellow of the Institute of Medicine of the National Academy of Sciences, the American Academy of Arts and Sciences, and the European Academy of Sciences and Arts. He lives in Los Angeles.